IEEE Press Series on
Computational Intelligence
www.wiley.com/go/cis

INTRODUCTION TO EVOLVABLE HARDWARE

IEEE Press
445 Hoes Lane
Piscataway, NJ 08854

IEEE Press Editorial Board
Mohamed E. El-Hawary, *Editor in Chief*

J. B. Anderson	S.V. Kartalopoulos	N. Schulz
R. J. Baker	M. Montrose	C. Singh
T. G. Croda	M. S. Newman	G. Zobrist
R. J. Herrick	F. M. B. Pereira	

Kenneth Moore, *Director of IEEE Book and Information Services (BIS)*
Catherine Faduska, *Senior Acquisitions Editor, IEEE Press*
Jeanne Audino, *Project Editor, IEEE Press*

IEEE Computational Intelligence Society, *Sponsor*
IEEE CI-S Liaison to IEEE Press, David B. Fogel

Technical Reviewers
Hugo de Garis, *Utah State University*
Jason Lohn, *NASA Ames Research Center*
Xin Yao, *The University of Birmingham, UK*
Ricardo S. Zebulum, *Jet Propulsion Laboratory*

Books in the IEEE Press Series on Computational Intelligence

Computational Intelligence: The Experts Speak
Edited by David B. Fogel and Charles J. Robinson
2003   0-471-27454-2

Handbook of Learning and Approximate Dynamic Programming
Edited by Jennie Si, Andrew G. Barto, Warren B. Powell, Donald Wunsch II
2004   0-471-66054-X

Computationally Intelligent Hybrid Systems
Edited by Seppo J. Ovaska
2005   0-471-47668-4

*Evolutionary Computation: Toward a New Philosophy of
Machine Intelligence, Third Edition*
David B. Fogel
2006   0-471-66951-2

Emergent Information Technologies and Enabling Policies for Counter-Terrorism
Edited by Robert L. Popp and John Yen
2006   0-471-77615-7

INTRODUCTION TO EVOLVABLE HARDWARE
A Practical Guide for Designing Self-Adaptive Systems

GARRISON W. GREENWOOD
Portland State University

ANDREW M. TYRRELL
University of York

IEEE Computational Intelligence Society, *Sponsor*

IEEE Press Series on Computational Intelligence
David B. Fogel, *Series Editor*

IEEE PRESS

A JOHN WILEY & SONS, INC., PUBLICATION

Copyright © 2007 by the Institute of Electrical and Electronics Engineers. All rights reserved.

Published by John Wiley & Sons, Inc., Hoboken, New Jersey.
Published simultaneously in Canada.

No part of this publication may be reproduced, stored in a retrieval system, or transmitted in any form or by any means, electronic, mechanical, photocopying, recording, scanning, or otherwise, except as permitted under Section 107 or 108 of the 1976 United States Copyright Act, without either the prior written permission of the Publisher, or authorization through payment of the appropriate per-copy fee to the Copyright Clearance Center, Inc., 222 Rosewood Drive, Danvers, MA 01923, (978) 750-8400, fax (978) 750-4470, or on the web at www.copyright.com. Requests to the Publisher for permission should be addressed to the Permissions Department, John Wiley & Sons, Inc., 111 River Street, Hoboken, NJ 07030, (201) 748-6011, fax (201) 748-6008, or online at http://www.wiley.com/go/permission.

Limit of Liability/Disclaimer of Warranty: While the publisher and author have used their best efforts in preparing this book, they make no representations or warranties with respect to the accuracy or completeness of the contents of this book and specifically disclaim any implied warranties of merchantability or fitness for a particular purpose. No warranty may be created or extended by sales representatives or written sales materials. The advice and strategies contained herein may not be suitable for your situation. You should consult with a professional where appropriate. Neither the publisher nor author shall be liable for any loss of profit or any other commercial damages, including but not limited to special, incidental, consequential, or other damages.

For general information on our other products and services or for technical support, please contact our Customer Care Department within the United States at (800) 762-2974, outside the United States at (317) 572-3993 or fax (317) 572-4002.

Wiley also publishes its books in a variety of electronic formats. Some content that appears in print may not be available in electronic formats. For more information about Wiley products, visit our web site at www.wiley.com.

Library of Congress Cataloging-in-Publication Data is available.

ISBN-13: 978-0-471-71977-9
ISBN-10: 0-471-71977-3

Printed in the United States of America.

10 9 8 7 6 5 4 3 2 1

*To my wife Linda
and my two children
Matthew and Sarah, who
all mean the world to me
—GWG*

*To my wife
Maggie for her everlasting
support and love
—AMT*

CONTENTS

PREFACE xi

ACKNOWLEDGMENTS xiii

ACRONYMS xv

1 INTRODUCTION 1

    1.1 Characteristics of Evolvable Circuits and Systems / 1
    1.2 Why Evolvable Hardware Is Good (and Bad!) / 5
    1.3 Technology / 6
    1.4 Evolvable Hardware vs. Evolved Hardware / 9
    1.5 Intrinsic vs. Extrinsic Evolution / 10
    1.6 Online vs. Offline Evolution / 12
    1.7 Evolvable Hardware Applications / 13
        References / 15

2 FUNDAMENTALS OF EVOLUTIONARY COMPUTATION 17

    2.1 What Is an EA? / 17
    2.2 Components of an EA / 18
        2.2.1 Representation / 18
        2.2.2 Variation / 21
        2.2.3 Evaluation / 23
        2.2.4 Selection / 26
        2.2.5 Population / 28

2.2.6   Termination Criteria / 29
2.3 Getting the EA to Work / 29
2.4 Which EA Is Best? / 31
     References / 32

3 RECONFIGURABLE DIGITAL DEVICES   35

3.1 Basic Architectures / 35
     3.1.1   Programmable Logic Devices / 38
     3.1.2   Field Programmable Gate Array / 39
3.2 Using Reconfigurable Hardware / 45
     3.2.1   Design Phase / 49
     3.2.2   Execution Phase / 50
3.3 Experimental Results / 53
3.4 Functional Overview of the POEtic Architecture / 55
     3.4.1   Organic Subsystem / 59
     3.4.2   Description of the Molecules / 59
     3.4.3   Description of the Routing Layer / 62
     3.4.4   Dynamic Routing / 62
3.5 Remarks / 64
     References / 64

4 RECONFIGURABLE ANALOG DEVICES   67

4.1 Basic Architectures / 67
4.2 Transistor Arrays / 70
     4.2.1   The NASA FTPA / 72
     4.2.2   The Heidelberg FPTA / 82
4.3 Analog Arrays / 89
4.4 Remarks / 93
     References / 93

5 PUTTING EVOLVABLE HARDWARE TO USE   95

5.1 Synthesis vs. Adaption / 95
5.2 Designing Self-Adaptive Systems / 96
     5.2.1   Fault Tolerant Systems / 96
     5.2.2   Real-Time Systems / 101
5.3 Creating Fault Tolerant Systems Using EHW / 102
5.4 Why Intrinsic Reconfiguration for Online Systems? / 103
5.5 Quantifying Intrinsic Reconfiguration Time / 104
5.6 Putting Theory Into Practice / 108
     5.6.1   Minimizing Risk With Anticipated Faults / 109
     5.6.2   Minimizing Risk With Unanticipated Faults / 111

       5.6.3   Suggested Practices / 113
  5.7 Examples of EHW-Based Fault Recovery / 113
       5.7.1   Population vs. Fitness-Based Designs / 114
       5.7.2   EHW Compensators / 117
       5.7.3   Robot Control / 125
       5.7.4   The POEtic Project / 138
       5.7.5   Embryo Development / 156
  5.8 Remarks / 172
       References / 172

6 FUTURE WORK   177

  6.1 Circuit Synthesis Topics / 177
       6.1.1   Digital Design / 178
       6.1.2   Analog Design / 184
  6.2 Circuit Adaption Topics / 185
       References / 187

INDEX   189
ABOUT THE AUTHORS   191

PREFACE

The complexity of electronic and computer systems continues to increase at a rate that one might consider too fast for designers to cope with. This increase in complexity enables us to produce impressive engineering systems: aircraft, cars, mobile phones, the internet, "intelligent" homes, for example. However, there are negative aspects to this increased complexity most obviously how to design, manage and reason about such complex systems. Biological systems are many orders of magnitude more complex than anything we can currently produce. Unfortunately as complexity increases so do the chances of faults and errors occurring in these systems. For non-maintainable systems, those that are not available to repair, such as satellites, deep-sea probes or long-range spacecraft, faults and errors can damage the system making the device useless. Are there different ways to produce our current and future complex engineering systems?

Evolvable hardware (EHW) is a dynamic field that brings together reconfigurable hardware, artificial intelligence, fault tolerance and autonomous systems. EHW uses simulated evolution to search for new hardware configurations. The evolution is performed by a variety of different stochastic search algorithms such as genetic algorithms, evolutionary programming or evolution strategies. The evolved hardware is implemented on reconfigurable devices such as field programmable gate arrays (FPGAs), field programmable analog arrays (FPAAs) or field programmable transistor arrays (FPTAs). Each device is configured to define its architecture (and thus function) and the purpose of this evolution is to find the best performing architecture for the given application.

EHW techniques have been successfully used for both original system design and online adaptation of existing systems. It is in latter application area that has generated the most interest. EHW allows systems to self-adapt to compensate for failures or unanticipated changes in the operational environment. This capability has attracted the attention of NASA and the military because operating in extreme physical environments, coupled with the need for high reliability, is the norm for systems deployed by these agencies. It is for this reason this book places a special emphasis on using EHW in fault tolerant applications.

This book provides a comprehensive view of this growing field for researchers, engineers, designers and managers involved in the design of adaptive and high reliability systems. The reader is introduced to the basic terminology and principles of EHW, reconfigurable hardware, algorithms that conduct the simulated evolution, and system integration concepts. Background information is included on real-time systems and fault-tolerant principles. Several real-world application examples (both digital and analog) are included to teach the basic concepts and to illustrate the power and versatility of EHW.

The motivation behind this book comes from the realization that all of the EHW literature is scattered in a variety of journal articles and conference proceedings. The authors have attempted to bring together, under one cover, the main concepts behind EHW, which will allow the reader to begin applying EHW techniques in a short period of time.

G. W. GREENWOOD

A. M. TYRRELL

ACKNOWLEDGMENTS

The authors wish to thank the following people for their help, whether implicitly or explicitly, in the production of this book: Cesar Ortega, Gordon Hollingworth, Renato Krohling, Will Barker, Crispin Cooper, Andy Greensted, Mic Lones, Alex Jackson, Richard Canham, Steve Smith, Gianluca Tempesti, Julian Miller, Yann Thoma, Manuel Moreno, Daniel Mange, Adrian Stoica, Ricardo Zebulum, David Fogel, Karlheinz Meier, and Joerg Langeheine. A particular thanks goes to Hugo de Garis, for his thorough review of the first draft of this book, and to Paul Garner, for his patience with the numerous changes to the figures!

A.M.T

G.W.G

ACRONYMS

ASIC	application specific integrated circuit
CLB	configurable logic block
CPLD	complex programmable logic device
COTS	commercial-off-the-shelf
CGP	cartesian genetic programming
DFT	design-for-test
EA	evolutionary algorithm
EDA	electronic design automation
EHW	evolvable hardware
FPAA	field programmable analog array
FPGA	field programmable gate array
FPTA	field programmable transistor array
FSM	finite state machine
HDL	hardware description language
OTP	one time programmable
PLD	programmable logic device

CHAPTER 1

INTRODUCTION

Aims: *We aim here to give a brief overview of evolvable hardware (EHW), to explain what it is and how it is used. Many more details will be given in subsequent chapters, but here we aim to give the reader a taste of things to come. We will also give some basic characteristics and examples of good and not so good features of evolvable systems.*

1.1 CHARACTERISTICS OF EVOLVABLE CIRCUITS AND SYSTEMS

In the last ten years the complexity of electronic and computer systems has increased dramatically. As more power is required more complex systems have been created to fulfill these demands. This increase in complexity enables us to produce, what to most of us are, impressive engineering systems: aircraft, cars, mobile phones, the internet, "intelligent" homes, to name just a few. However, there are potentially negative aspects to this increased complexity. Most obviously how to design, manage and reason about such complex systems. It is interesting to make the point now that biological systems are many orders of magnitude more complex than anything we can currently produce. In addition, unfortunately as complexity increases so do the faults and errors seen in these systems. For non-maintainable systems, those that are not available to repair due to the expense, such as satellites, deep-sea probes or long-range spacecraft, faults and errors can damage the system making the device useless.

Introduction to Evolvable Hardware: A Practical Guide for Designing Self-Adaptive Systems, by Garrison W. Greenwood and Andrew M. Tyrrell
Copyright © 2007 Institute of Electrical and Electronics Engineers

Let's consider in a little more detail here the ideas behind faults in systems; we will talk much more about this in later chapters. Faults in high integrity computer system can be detected and fixed through maintenance. Fault tolerant techniques are provided to allow the system to continue working even when one or more of its redundant parts have failed. In non-maintainable systems this is not an option; once faults have occurred and been masked through fault tolerance techniques, the specific element of the system is no longer used and it cannot be repaired to be used again. Once all of the redundant elements have been destroyed through different faults the system will fail.

One extreme example of this is the one hundred year mission to the nearest star. Over the course of one hundred years an electronic system will have degraded to the point where very little of the original hardware will still be operational. In this case, if the system fails, the whole mission will fail. There is no chance of sending out maintenance engineers to fix the craft.

Consider Figure 1.1, a rather stylized model of the evolutionary cycle that occurs for all life. Details of this cycle as applied to engineering systems, and particularly evolvable hardware are given in later chapters. But for now let's consider this cycle. We will start with parents (we have to start somewhere and we don't wish to get into the chicken and egg argument!). These parents will produce children, who will exist in an environment to which they are better or worse suited than others. Selection will occur to determine if these children will (or will not) become parents for the next generation.

The important points here are that we have a population of parents, who produce a new population of children that form the next generation. Some form

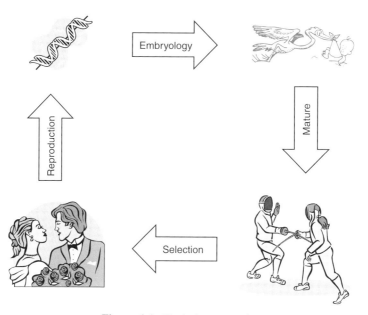

Figure 1.1. Evolutionary cycle.

of selection will take place and success in this selection process will enable members of the population to participate in the creation of the next generation (non-selection means obscurity for eternity!)

How might we now move this towards our evolvable hardware system? Consider Figure 1.2. Our population is now a set of electronic systems attempting to solve the same problem. If we assume that we can transform the problem solution into a measurable number—called "fitness" in Figure 1.2—we can use this number to help us with the selection stage of our cycle. Again, those selected will be involved in forming the next generation of possible solutions to the problem. DNA from Figure 1.1 has now been transformed into a binary string (it does not have to be a binary string, but at this early stage in our discussions this is a good example). This binary string is used to "configure" our system. As illustrated in Figure 1.2, this cycle continues, maybe forever, or more typically until the fitness achieves its maximum value or until we get bored!

Figure 1.3 illustrates this process again. This time its representation is more closely aligned to electronic hardware. It also illustrates some of the basic ideas used in the field to introduce some randomness into the whole process: In this case crossover and mutation (again much more will be said about these in Chapter 2).

These are the basics of evolvable hardware. Of course, there are many details we have not mentioned here, but these will be covered in later chapters. This quick overview is intended to just cover the underlying ideas behind the subject.

What are the characteristics that one might expect to find in an evolvable hardware system? It is difficult to generalize because the answer depends on a

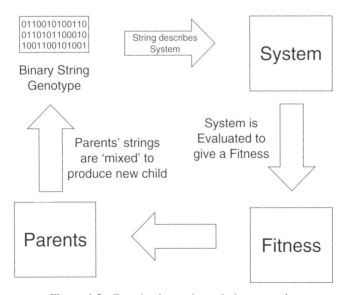

Figure 1.2. Generic electronic evolutionary cycle.

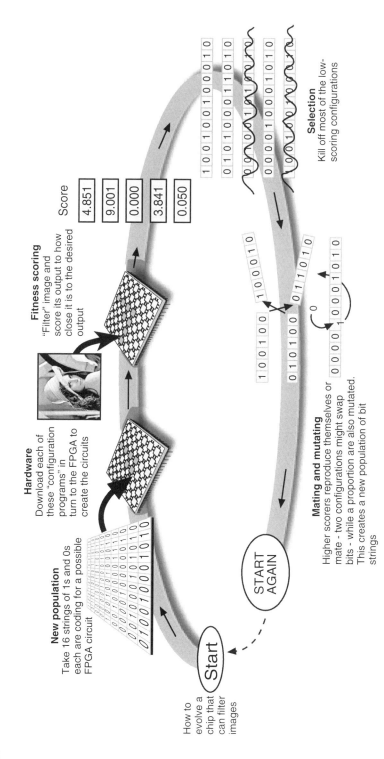

Figure 1.3. Hardware evolutionary cycle.

number of factors decided at the beginning of the process. However, here are a few:

- The system will be evolved not designed (an obvious one, but worth pointing out!).
- It may not be an optimal design, but will be fit for the purpose (typical of biological systems in general—due to the evolutionary process).
- Evolved systems do show levels of fault tolerance not seen in designed systems (again typical of biological systems).
- Adaptability to environmental changes (this is dependent upon when the evolutionary cycle stops. This is not necessarily true for evolved systems that stop once the goal has been reached. Again this is a characteristic of all biological systems).
- Unverifiable systems are produced, indeed in many cases it is difficult to analyze the final system to see how it performs the task!

There are more for sure, but as we have already mentioned application dependent. For now we will leave it, but you will see many more examples of evolvable hardware as you go through the book, and hopefully you will try some for yourself, by doing this you will experience your own trials and tribulations, excitement and frustrations, but will probably come up with your own set of characteristics!

1.2 WHY EVOLVABLE HARDWARE IS GOOD (AND BAD!)

As we have seen, evolvable hardware is a method for electronic circuit design that uses inspiration from biology to evolve rather than design hardware. At first sight this would therefore seem very appealing. Consider two different systems, one from Nature and one human engineered: the human immune system (Nature design) and the computer that we are writing this book on (human design). Which is more complex? Which is most reliable? Which is optimized most? Which is most adaptable to change?

The answer to almost all of those questions is, I hope you will agree, the immune system. About the only winner from the engineered side might be the question "Which is optimized most?" So the obvious conclusion might be, let's get rid of all of our design methods and let's do it the way Nature does! Well as you can probably guess it is not quite that straight forward as we will see throughout this book. Nature has a number of advantages over current engineered systems, not least of which are time (most biological systems have been around for quite a long time!) and resources (most biological systems can produce new resources as they need them, e.g. new cells).

However, this does not mean that evolutionary designs can't be useful (we would not be writing this book if we thought that!) We hope to show in this book where and when evolutionary designs are advantageous and how best to make use of them. For now just to give one example where evolutionary techniques can have a real benefit, and to help us persuade you to keep reading. Bear in mind one of the primary goals is to keep our electronic systems up and running for long periods of time, often in unpredictable environments.

Since the basis of evolutionary algorithms (see Chapter 2) is a population of different systems which compete for the chance to reproduce, it already contains a type of redundancy, since each system in the population is different. When a fault occurs in the system, either through external conditions (sensors failing or giving incorrect readings), or through internal electronic faults, it will affect different systems in different ways due to the diversity of solutions. This should mean that particular individuals will not be effected by the fault and therefore provide tolerance to the fault.

For example, Figure 1.4 [1] illustrates the results of a successful evolutionary process which was trying to design a simple frequency generator on a Field Programmable Gate Array (FPGA). The details are not important. What is interesting is to see the different patterns. Each light block is a functional unit on the FPGA that is performing part of the function in the process (the dark blocks are not being used). This figure shows eight members of a population, all produced by the same evolutionary process. What we can see is that different sets of functional units are used in each member, even though they all do the job effectively. Hence a fault in one part of the FPGA may effect the performance of one circuit, but not another—hence a more resilient design is produced.

This is one very simple, but important, example of where evolutionary techniques might have an advantage over more traditional designs. We will show others through the pages of this book.

1.3 TECHNOLOGY

The field—of evolvable hardware has been driven by—in some circumstances, waiting on–technology. Many of the things we are going to discuss in this book are only possible because of the underlying technology they are based on. Many of the ideas, such as systems that evolve, have been around for a number of decades. However, actual devices to enable us to test such ideas on real hardware have only been around for less than a decade. Hence, we feel it is worth saying a few words about the technology that evolvable hardware relies on.

First, let's consider size. Where are we in technology compared to other aspects of the world around us and where are we going? A ball used in football—or soccer if you prefer—is about 22 cm in diameter. A human hair around 80 μm and a blood cell 7 μm. A strand of DNA about 2 nm wide. A comparison of these and other elements are shown in Figure 1.5.

In microelectronics, 10 years ago we where fabricating devices with feature size of around 2–3 μm. In 2004 it is around 0.09 μm. What is the effect of this reduction in size, and will this general reduction in feature size continue?

The effects of this reduced feature size is, at least simply, that we can get more features onto the same size chip (device). It also means we can probably make

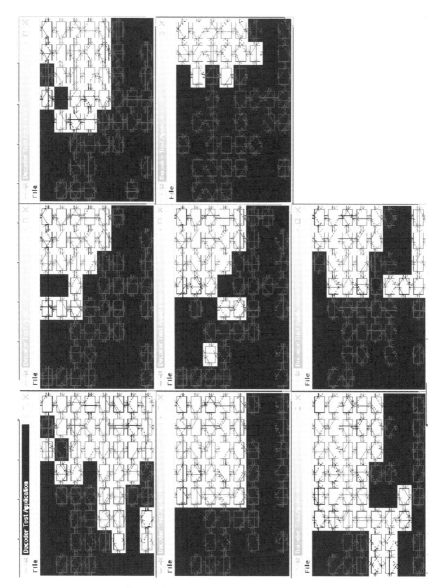

Figure 1.4. Result of an evolved design.

things go faster (i.e., a faster clock speed). The features are smaller so switch quicker, they are closer together, because they are smaller, so transmission of signals is quicker. So, to a first order of approximation we have more functionality and quicker devices.

What about the trend and will it continue? Gordon Moore from Intel back in the 1960s observed that the number of transistors on silicon devices doubled

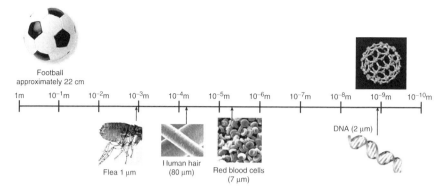

Figure 1.5. Size is everything! [2].

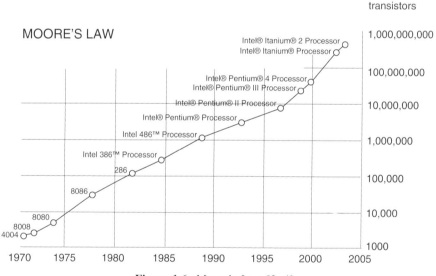

Figure 1.6. Moore's Law [3, 4].

every 12–18 months. He then made, at the time what might have been a rather rash prediction, that the number of transistors would continue to double every 18 months. This prediction has proved to be true (whether because this is the natural rate for this technology or because all IC manufacturers now have this as their roadmap and therefore are almost obliged to meet it?!) and is illustrated in Figure 1.6.

Will it continue? Probably not. We are approaching a time in the next 10–15 years when the feature size will be down to the atomic level and traditional techniques will not allow us to go smaller. Then we might be looking at Quantum Computing, Molecular Computing, DNA Computing or something else. However, we think for now we can stop here!

How does this affect our area of evolvable hardware? It is the extra functionality that interests us more than the speed at the moment (although extra speed is always nice). The reduction in feature size, and the subsequent increase in functionality, has meant that for the first time device vendors can put sufficient resources on a single chip to place memory next to processing. In other words, on-chip programmability is now possible. It also allows enough on-chip transport medium to have a significant amount of communication lines (buses) within a single chip. All of this (and more) is critical if we are to perform real evolvable hardware. Much of this needed capability has only been available in the last 10 years, which explains why the growth in this exciting field is only now taking place.

1.4 EVOLVABLE HARDWARE VS. EVOLVED HARDWARE

What do we mean by "evolvable hardware" and "evolved hardware"? The two terms are not interchangeable.

Evolvable hardware is a new field that brings together reconfigurable hardware, artificial intelligence, fault tolerance and autonomous systems. Evolvable hardware refers to hardware that can change its architecture and behavior dynamically autonomously by interacting with its environment. Ideally this process of interactions and changes should be a continuous one and should be open-ended.

Conversely, evolved hardware refers to hardware that has been created through a process of continued refinement and stops when a sufficiently "good" individual has been found.

On first reading you may think that these two definitions are the same, but they are not. Both use evolutionary techniques to produce a system that performs to some specification, both are to some extent autonomous, and both might have properties that give the final system some fault tolerance. The major difference between the two is that evolved systems stop changing after a good individual has been found and thereafter remain static designs (as are most human designs). Evolvable systems continue to have the possibility of changing throughout their existence. Hence, as systems change (e.g. components fail), as the environment changes (temperature changes, sensor failures, etc.) an evolvable system will be able to adapt to these changes and continue to operate effectively. Again, let's consider a simple example, this time based around a small robot.

Figure 1.7 shows the Khepera robot, which was used in a series of very interesting experiments. The robot has eight infrared proximity sensors and two wheels. Each sensor can emit infrared light and detect the reflected signals. The sensor value is varied from 0–1023. The higher the sensor values the closer the distance between sensor and obstacle. Each wheel of the robot is controlled by an independent DC motor. The controller receives 2 bit information for four commands: move forward, move backward, turn left and turn right. The controller was implemented on an FPGA within a PC (the PC provided power and

10 INTRODUCTION

Figure 1.7. The Khepera robot and its experimental test environment.

a programming environment for the FPGA, it did not take part in the evolutionary processes). The task was simply to avoid objects. Evolution was continuous, hence changes could occur even after the best of a population was chosen.

Results are presented here for the case where faults were added to some of the sensors. These faults were added by covering one of the robot sensors with a paper mask, which fixes the sensor output. A fault was added after an acceptable controller had already been evolved—that is, during run-time.

A fault was added to sensor 2 at generation 50. The results are shown in Figure 1.8. This fault disrupted the controllers of the whole population. Observe how both the average fitness and the fitness of the best individual dropped when the fault was added. However, after around 10 generations the fitness of the best individual climbs up and regains its original value. Figure 1.8 shows that although this approach does not prevent the failure introduced by the fault (drop off of fitness) when it occurs, it does allow it to regain the fitness quite quickly.

This example shows a real advantage of evolvable designs over evolved or more traditional designs. It can react to its environment and other internal states, in this particular case a sensor failure, and adapt and recover from this and continue to provide an appropriate response.

1.5 INTRINSIC VS. EXTRINSIC EVOLUTION

Figure 1.9 shows the main steps in the evolutionary synthesis of electronic circuits. First a population of chromosomes—i.e., encoded circuit configurations—is randomly generated. Each bit in the chromosome defines some architectural feature

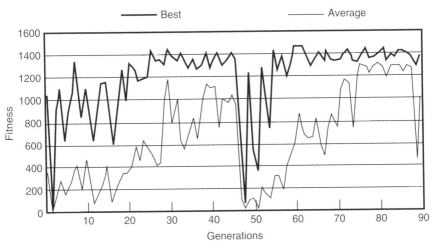

Figure 1.8. Behavior for the best and average fitness of the population when a fault in sensor 2 of the robot is added at generation 50. (See [5]).

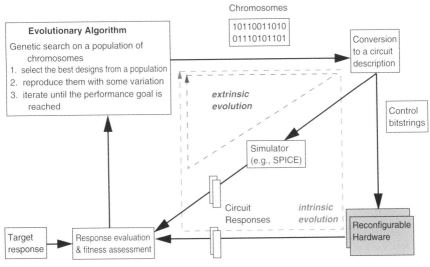

Figure 1.9. Evolutionary synthesis of electronic circuits.

of the reconfigurable hardware. For instance, a bit might give the state of a switch that connects two circuit components. Second, every candidate configuration must be checked for "fitness", which measures how closely it matches a target response. Third, highly fit chromosomes, which match the target response quite well, are subjected to stochastic genetic operators that create new chromosomes for evaluation. This process is repeated for numerous generations, resulting in better fit configurations. Normally the best fit configuration from the last generation is the one chosen by the designer.

There are two different ways of checking fitness, but which method is used depends on whether the evolution is done *extrinsically* or *intrinsically* Extrinsic evolution uses circuit models and simulators to evaluate circuit configurations. The evaluation is done entirely in software and only the one best configuration from the last generation is converted to control bits and downloaded to program the reconfigurable hardware. On the other hand, in an intrinsic evolution every chromosome is downloaded and physical testing measures fitness. Put another way, extrinsic evolution is software evolution whereas intrinsic evolution is hardware evolution. There are advantages and disadvantages to each method of evolution. A detailed discussion is in Chapter 5.

1.6 ONLINE VS. OFFLINE EVOLUTION

EHW methods can be used for creating digital, analog or mix-signal circuitry. Applications fit into two broad categories:

1) *Circuit Synthesis*

The objective here is to design a circuit that satisfies a set of functional and timing specifications. Complete information about the physical operating environment is also available. There is no preliminary or prototype hardware in existence. Essentially, the design has to be done from scratch.

Synthesis problems have three things in common which EHW techniques can exploit. First, there are virtually no restrictions on computing resources. The evolutionary algorithm can run on anything ranging from a laptop computer to massively parallel multiprocessor systems. Memory shortage is of no concern. Second, either extrinsic evolution or intrinsic evolution is possible. Physical measurements using sophisticated digital storage oscilloscopes or spectrum analyzers can be conducted. Moreover, testing in extreme environments such as high radiation or ultra-low temperatures is possible. Third, EHW-based synthesis problems are solved in laboratory environments. Under these conditions there are few restrictions on simulation times or the number of generations an evolutionary algorithm can run. Exhaustive testing can also be conducted.

Fitness in circuit synthesis problems is always precisely defined, often with closed-form expressions. EHW methods used for synthesis problems are run offline.

2) *Adaptive Hardware*

Hardware is always adapted *in-situ*[1]. This type of operational environment presents a number of daunting challenges for EHW practitioners: (1) the evolutionary algorithm has to run on limited computing resources, such as a low-end

---
[1] A latin term meaning in position or in place.

microcontroller with limited memory; (2) the evolutionary algorithm often has a pre-defined maximum running time; (3) no laboratory instruments are available to test evolving hardware; and (4) the evolution is performed with little or no human intervention or oversight.

EHW techniques that adapt hardware are completely different from those used to perform synthesis—in part because they are needed for a different purpose. Existing hardware is adapted to compensate for a changed operational environment, to correct a fault, or to compensate for aging effects. Frequently the evolution will be done with incomplete information about how the operational environment changed or without knowing the true nature of the fault. Consequently, fitness is seldom precisely defined and almost never with a closed form expression. Intrinsic evolution is the norm. EHW methods for hardware adaption problems are exclusively run online while the hardware operates in a real physical environment.

1.7 EVOLVABLE HARDWARE APPLICATIONS

EHW-based circuit synthesis techniques are alternatives to the conventional design methods used today—including hand-design methods. This means EHW-based methods must provide a clear advantage over existing design methods if EHW is to gain any footholds. Industry today has very powerful, computer aided methods for synthesizing digital circuitry, thereby creating a steep hill to climb for any new synthesis method. On the other hand tools for analog synthesis are largely nonexistent despite a growing interest in mixed-signal applications. Very powerful analog simulators such as PSpice are on the market and they have been of enormous benefit to both analog engineers and EHW researchers. Analog synthesizers are still not yet widely available—making this an area where EHW-based techniques can have a real impact.

Unfortunately, scaling problems are still not resolved. Most evolved circuitry has a small number of components compared to what conventional methods can produce. (This issue is discussed further in Chapter 5.) But the real promise of EHW lies with circuit adaptation, where a circuit autonomously improves its behavior while operating in a real physical environment [6].

The reader should proceed here with caution, for no matter how seductive this capability might sound, there are some strong caveats that make this sometimes difficult—maybe even undesirable—to achieve. For one thing reconfigurable hardware is never used as a stand-alone system; it is always an embedded system with physical interfaces to other systems. Hence, any modification in its behavior will likewise affect the behavior of other systems, whether they are reconfigurable or not. Although evolution can, under some circumstance, lead to improved circuit behavior, this improvement is not monotonic. Along the way there are evolved circuits that exhibit worse behavior. Since the reconfigurable hardware doesn't operate in isolation it doesn't adapt in isolation either. Therein lies the crux of the problem with online adaption. What do you do with the rest of the system while one of its subsystems is reconfiguring? Designers cannot ignore the potential risk reconfiguration might have on any interfaced systems. In the worse case the larger system will have to be taken offline to prevent any damage during reconfiguration operations.

There are two situations where reconfiguration becomes necessary. The first situation is where the operating environment changes. Deep-space probes are a good example because they can encounter unknown environments where temperatures fluctuate rapidly and high-radiation zones pose a real threat because they change the performance characteristics of electronic components. Reconfiguration provides a realistic way of compensating for any component changes. But component values also change over time even if the operating environment doesn't change. This drift in component properties is referred to as *aging effects* and it is a serious threat to deep-space probes that are expected to operate for decades. Reconfiguration can help mitigate aging effects.

The second situation where reconfiguration can be immensely beneficial is in fault tolerant systems. All systems eventually fail but timely repair or replacement is impractical or even impossible. This is especially true for autonomous systems that must operate unsupervised in harsh environments for extended periods of time without support. Deep-space probes and unmanned aerial vehicles (UAVs) are two obvious examples. Such systems have limited space and severe weight restrictions that don't allow for spare hardware so reconfiguration may be the only viable means of fault recovery. Nevertheless, autonomous fault recovery under any circumstances is not a trivial task. Fault recovery via reconfiguration is the focus of Chapter 5.

EHW-based techniques have been around since the early 1990s. Designers now have alternative methods for synthesizing circuits and new ways of addressing fault tolerant problems. But along with those advantages are also some significant difficulties. Indeed, a number of fundamental issues including scalability and real-time, online adaptation remain open. Much work remains to be done. This book should help the reader understand the promises and challenges surrounding EHW.

We do not wish to leave the reader with the impression that the sole purpose of EHW-based techniques is to create fault tolerant architecture. Recently some very interesting work in embryonic hardware has shown that, in addition to fault tolerance, EHW methods can create very complex and robust circuitry (e.g., see [7]). Chapter 5 will discuss embryonic hardware in greater detail.

Before starting, however, it is important to understand the purpose of EHW is to evolve circuit *behaviors* and not circuit *structures*. An essential part of any EHW technique is the evolutionary algorithm that searches for a good circuit configurations. This evolutionary algorithm relies on a fitness function to evaluate a circuit. But what is evaluated? Certainly not whether individual components are interconnected in some specified order. Instead the circuit is evaluated to see if it has the right frequency response or perhaps some specific timing. These are behavioral properties not structural problems.

This distinction is not just a philosophical interpretation; it goes to the heart of what EHW is all about. Circuit design always begins with a specification.

Remember that the purpose of a design specification is to state *what* needs to be designed and not *how* to do that design. In other words, the specification tells the designer what performance is required—it is up to the creativity of the designer to come up with a circuit that meets those requirements. Hence, EHW is a design method more in line with the way designers actual do their job.

REFERENCES

1. Hollingworth G, Smith S. and Tyrrell A 2000, "The Intrinsic Evolution of Virtex Devices through Internet Reconfigurable Logic", *3rd International Conference on Evolvable Systems: from Biology to Hardware, Edinburgh*, Lecture Notes in Computer Science, Springer-Verlag, Heidelberg, 72–79
2. Nanoscience and Nanotechnologies: opportunities and uncertainties http://www.nanotec.org.uk/finalReport.htm
3. http://www.intel.com/research/silicon/mooreslaw.htm
4. Moore G 1965. "Cramming more components onto integrated circuits", *Electronics* 38, 114–118
5. Tyrrell A, Krohling R and Zhou Y 2004, "A new evolutionary algorithm for the promotion of evolvable hardware", *IEE Proceedings of Computers and Digital Techniques* 151(4), 267–275
6. Yao X and Higuchi T 1999, "Promises and challenges of evolvable hardware", *IEEE Transactions on Systems, Man & Cybernetics—Part C* 29(1), 87–97
7. Mange D, Sipper M, Stauffer A and Tempesti G 2000, "Towards robust integrated circuits: the embryonics approach", *Proceedings of the IEEE* 88(4), 516–541

CHAPTER 2

FUNDAMENTALS OF EVOLUTIONARY COMPUTATION

Aims: *This chapter describes what an evolutionary algorithm (EA) is, what its major components are, and how the algorithm is used. Only an overview is presented so all aspects of EAs cannot possibly be discussed. Nevertheless, the material provides sufficient preparation so readers can begin to study the EHW literature. More detailed information about EAs is available in a number of recently published books. The Eiben and Smith [3] book is particularly good.*

2.1 WHAT IS AN EA?

Evolutionary algorithms are computer algorithms that mimic the forces of natural evolution and self-adaptation to solve difficult problems. It is therefore not surprising that the underlying theory and even the terminology has strong ties to evolutionary biology. More specifically, EAs follow the neo-Darwinian philosophy which says stochastic processes such as reproduction and selection, acting on species, are responsible for the present life forms we know. In simple terms natural evolution describes how a population of individuals strives for survival. During reproduction genetic material from each parent creates an offspring. Each individual has an associated fitness that ultimately determines the survival probability. Highly fit individuals have a high probability of surviving to reproduce in future generations.

With respect to EAs, each individual is a unique solution to the optimization problem of interest. A population of individuals is therefore a set of possible

---

Introduction to Evolvable Hardware: A Practical Guide for Designing Self-Adaptive Systems,
by Garrison W. Greenwood and Andrew M. Tyrrell
Copyright © 2007 Institute of Electrical and Electronics Engineers

solutions and the fitness of a solution measures its quality. New individuals are created each generation by randomly varying individuals in the current population. Every individual is evaluated and the best individuals are selected for the next population. That is,

$$P(t+1) = \mathcal{S}(\mathcal{E}(\mathcal{V}(P(t)))), \tag{2.1}$$

where $P(t)$ is a population of solutions in generation t, $\mathcal{V}(\cdot)$ is a random variation operator, $\mathcal{E}(\cdot)$ is an evaluation operator, and $\mathcal{S}(\cdot)$ is a selection operator. This process repeats every generation until the termination criterion is satisfied. Each component of an EA is described in the next section.

2.2 COMPONENTS OF AN EA

2.2.1 Representation

Representation refers to the data structure that encodes all the problem parameters needed to describe a solution. Biological terms are often used when describing representations. In the biological world the term *genome* refers to all of the genetic material that can be used to build a given life form, but only a subset of this material, called the *genotype*, is needed to build any particular individual. Each gene value or *allele* has a defined location or *locus* within the genotype. Evolution therefore alters one genotype to create a different genotype by changing an allele at a specified locus. The *phenotype* is the physical realization of the genotype. The phenotype tells the genotype's behavior in a specific environment.

Each encoded problem parameter in an EA is considered a *gene*. Hence, in EHW problems the genome is the set of all possible encoded problem parameters, and the genotype is the subset of those parameters needed to describe a particular circuit configuration[1]. The phenotype is the actual problem solution (i.e., a physical circuit).

An example will help clarify the relationship between a genotype and a phenotype with respect to EHW problems. Suppose you want to evolve a passive lowpass filter that satisfies a given specification. To build a lowpass filter you have to know how many resistors, capacitors and inductors to use, their values and possibly how to connect them together. The genotype furnishes that construction information while the phenotype is the filter circuit built from that information. Put another way, the genotype tells you how to build the circuit and the associated phenotype is the circuit itself. Note that the phenotype could be either a computer model or a real circuit built with physical components.

There are five encodings commonly used in EHW problems:

1. Binary strings
2. Integers

---

[1]The term chromosome is often used in lieu of genotype if the representation is a binary string.

3. Real numbers
4. Graphs
5. Hybrids

2.2.1.1 Binary Strings Binary strings are arguably the most frequently encountered encodings. They are very versatile because they can encode both component values and the circuit topology (configuration). Consider an RC network with possible resistor values $R = \{100\ \Omega,\ 220\ \Omega,\ 390\ \Omega,\ 470\ \Omega,\ 1\ K\Omega,\ 2.2\ K\Omega,\ 3.3\ K\Omega,\ 5.6\ K\Omega\}$ and possible capacitor values $C = \{10\ \mu F,\ 47\ \mu F,\ 220\ \mu F,\ 1000\ \mu F\}$. Any circuit configuration can be encoded with a 6-bit binary string: three bits to select one of eight R values (000 for 100 Ω, 001 for 220 Ω and so on); two bits to select one of four C values (00 for 10 μF, 01 for 47 μF and so on); and one bit to indicate if the R and C are in series (logic 0) or parallel (logic 1). For instance, the binary string 011100 would indicate a 470 Ω resistor in series with a 220 μF capacitor.

2.2.1.2 Integers In some cases it is more convenient to specify component values directly. The integer array is split into fields where each field is associated with a particular component. The units are predefined so only the numerical value is encoded. For example, the integer array

470	220

could describe a series RC network where the units were predefined as ohms and microfarads. The integer representation is easy to decipher the component values—which is particularly helpful if the circuit has a large number of components. However, it is not easy to describe circuit topologies with this representation.

2.2.1.3 Real Numbers Real numbers are needed to represent coefficient values such as tap weights for a digital filter or gain constants for a PID controller. An ℓ-bit binary string is a compact representation for a real number $x \in [x_{min}, x_{max}]$. Converting the binary string into a real number is straightforward. First, the binary string $b_{\ell-1} \ldots b_1 b_0$ is transformed into an (unsigned) integer $\tilde{x}$

$$\tilde{x} = \sum_{i=0}^{\ell-1} b_i \cdot 2^i \quad : \quad b_i \in \{0, 1\}$$

Second, $\tilde{x}$ is substituted into

$$x = x_{min} + \tilde{x} \cdot \frac{x_{max} - x_{min}}{2^\ell - 1}$$

to get x. One additional bit can be added for signed numbers (0 for positive, 1 for negative). The designer picks ℓ to give a desired precision level. Normally $\ell = 10$ or 12 is more than sufficient for EHW problems.

The 32-bit version of the IEEE-754 Standard is also widely used to represent real numbers, although it provides far more precision than what is really needed for EAs in general and EHW problems in particular—signed numbers from 2^{-128} to 2^{127} can be represented! Real number variables are often collected into a vector.

2.2.1.4 Graphs By definition, a graph is a set of vertices connected by a set of edges. A *directed graph* (or digraph for short) has direction associated with the edges. The graph is *undirected* if the edges have no direction. Figure 2.1 shows two example graphs.

Let v and v' be two vertices. A path between v and v' is an alternating sequence of vertices and edges without any repeats except possibly v or v'. The path is a *cycle* if $v = v'$—i.e., the path begins and ends at the same vertex. The graph is *connected* if there is a path between every pair of vertices. A connected graph containing no cycles is a *tree*.

Analog circuit topologies are naturally encoded as trees by interpreting vertices as components (resistors, transistors, etc.) and edges as interconnections. Digraphs are better suited for encoding digital circuit topologies because logic device input and outputs, other than data, are normally unidirectional. Digraphs are particularly useful for encoding array topologies. Neural networks are a good example.

2.2.1.5 Hybrids These encodings use a mixture of different encodings. Hybrid representations are particularly useful when both circuit topologies and component values are evolved. For instance, integer fields could be used to specify component values while binary fields could encode switch positions.

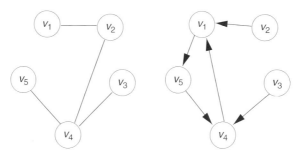

Figure 2.1. Two example graphs where the circles are vertices and the lines between them are edges. The left graph is undirected while the right graph is directed. Only the left graph is a tree. (See text.).

2.2.2 Variation

Variation is a random process that creates offspring (new individuals) from parents (existing individuals) by changing some or all of the encoded solution parameters. The most common variation operators are *mutation* and *recombination*.

2.2.2.1 Mutation Mutation requires only one parent. The idea is to create a single offspring by randomly altering one or more encoded solution parameters in the parent. (In practice the parent itself is not mutated but rather a copy of the parent.) The exact mutation operator form depends on the individual's representation. For example, a simple mutation operator for individuals encoded as a binary string is to complement a randomly chosen bit position[2]. Of course more than one bit position may be complemented. A common practice is to complement every bit position with some small probability p_m (typically less than 0.01).

More extensive mutations are possible. Entire blocks of consecutive bit positions can be swapped. This swap operator can also be used with more complicated representations. For instance, if the individual is encoded as an integer array, two randomly chosen integers can be swapped

    parent: 1 2 **5** 8 7 3 **6** 4 0
    offspring: 1 2 **4** 8 7 3 **6** 5 0

Another mutation method is *inversion* where the order of all integers between two randomly chosen locations are reversed.

    parent: 1 2 5 **8 7 3 6** 4 0
    offspring: 1 2 5 **6 3 7 8** 4 0

A very effective way for mutating real number parameters is to add a small random variable to the current parameter value. The magnitude of this random variable determines the mutation strength. Let $x_i \in \Re$ for $i = 1 \ldots n$ and let $N(0, 1)$ denote a normally distributed random variable with zero mean and unity variance. Then x_i, which is called an *object parameter*, is mutated by

$$x'_i = x_i + \sigma \cdot N_i(0, 1) \qquad (2.2)$$

where σ is the mutation step size and $N_i(\cdot, \cdot)$ means a new random variable is sampled for each x_i. σ is called a *strategy parameter*. Studies have shown that evolving the strategy parameters along with the object parameters improves the quality of the search [13]. The strategy parameter is adjusted online without using any external, deterministic controls (a process called *self-adaptation*). There are many ways of doing self-adaptation. One of the easiest methods is

$$\sigma'_i = \sigma_i \cdot \exp[\tau \cdot N_i(0, 1)] \qquad (2.3)$$
$$x'_i = \sigma'_i \cdot N_i(0, 1)$$

---
[2]In the literature "complementing a bit" is sometimes called "flipping a bit".

where τ is a user-defined parameter (normally a function of n). Notice that each x_i has its own unique step size.

2.2.2.2 Recombination Recombination uses genetic material from two or more parents to create offspring. As with mutation, the recombination operator depends on the representation. The simplest recombination operator is 1-point crossover. Consider a binary string of length ℓ. Two parents are split at bit position k and all bits after bit position k are exchanged. Figure 2.2 shows an example of 1-point crossover.

This concept can be applied to the more general n-point crossover where n random bit positions are randomly chosen. With n-point crossover alternate segments from each parent are used to form the offspring. See Figure 2.3.

Uniform crossover creates one offspring from two parents by treating each bit independently. For each bit position a random number over [0,1) is chosen. If the random number is less than 0.5 than the bit value is copied from the first parent. Otherwise the bit value from the second parent is copied. See Figure 2.4.

Recombination can also be done with real number vectors. The crossover operators described above—which are typically called discrete recombination operators—can be used with real number vectors, but that is not the only possibility. For example, with intermediate recombination one offspring is created from two parents by averaging the parent's component values. In fact, one can even do panmictic recombination where three or more parents participate in the component-wise averaging.

Up to this point we have not restricted how a variation operator can change a genotype. This is not always the case. Constrained optimization problems put certain limits on what constitutes a legal solution. For example, reconfigurable analog devices implement different circuit configurations with programmable switches.

Figure 2.2. Example of 1-point crossover. The line indicates the crossover point in the parents. Notice that the bits after the crossover point are swapped in the offspring.

Figure 2.3. Example of 2-point crossover. The bits between the two crossover points are swapped in the offspring. All other bits are the same as in the parents.

r.v = {0.6, 0.1, 0.3, 0.7, 0.4}

Figure 2.4. Example of uniform crossover. Each bit has an associated random variable. If the random variable value is greater than 0.5, the bit from the upper parent is used in the offspring. Otherwise the bit from the lower parent is used.

Although any switch can be opened, opening all the switches may isolate the inputs, which makes the circuit inoperable. Any legal solution is therefore prohibited from having all switches programmed open. Constraints such as this can be incorporated into the variation operator design itself—i.e., fix certain alleles and design the operator so that they are never modified. It may also be possible to repair the genotype by forcibly changing any improper alleles. Either way, the variation operator will always produce legal offspring from legal parents.

2.2.3 Evaluation

Evaluation is done using an *objective function* that assigns a numeric score to a solution, thereby indicating its quality. Normally this objective function maps the solution onto the real number line. That is, if $\mathcal{O}$ is an objective function and X is the space of all possible solutions, then $\mathcal{O} : X \rightarrow \Re$. Solution a is considered better than solution b if $\mathcal{O}(a) > \mathcal{O}(b)$.

The objective function should assign a numeric score that accurately reflects how closely a circuit meets the design specifications. All design parameters of interest should contribute to this score. This means the objective function is tailored to the optimization problem being solved. An example will illustrate the idea. Suppose we are using an EA to find appropriate values for the passive components in a low-pass filter. The design specification would describe the filter's desired spectral properties. For example, the specification might say the filter must have 0 dB gain at frequency $F_1 = 100$ Hz, -3 dB gain at frequency $F_2 = 200$ Hz, and -25 dB gain at frequency $F_3 = 500$ Hz. Assume gain values were specified for M different frequencies. One possible objective function is

$$\mathcal{O}(s) = \sum_{i=1}^{M} [G'(i) - G(i)]^2, \qquad (2.4)$$

where s is a filter circuit in the solution space X, $G(i)$ is the gain of a filter s at frequency F_i and $G'(i)$ is the ideal gain at frequency F_i. Filter s is better than filter s'—i.e., filter s is a closer match to the design specifications—if $\mathcal{O}(s) > \mathcal{O}(s')$. The goal is to search for the optimal filter s^* where $\mathcal{O}(s^*) \geq \mathcal{O}(s) \forall s \in X$.

Sometimes the literature refers to a *fitness function* rather than to an objective function. The term "fitness" is arguably more appropriate for EAs since they are based (loosely) on neo-Darwinian principles. However, the terms "fitness function" and "objective function" are not necessarily interchangeable. Fitness implies health and suitability—attributes that should always be maximized—whereas objectives could be minimized or maximized depending on their definition. The objective function defined in Eq. (2.4) is to be minimized. However, if the definition was

$$\mathcal{O}(s) = \frac{1}{\sum_{i=1}^{M}[G'(i) - G(i)]^2 + \varepsilon} \quad ; \; \varepsilon \ll 1, \quad (2.5)$$

then this objective should be maximized—which fits nicely with the notion of fitness. (ε is needed to keep $\mathcal{O}(s)$ finite.) In practice either fitness function or objective function could be used, although technically fitness function is really only appropriate if maximization is desired. Objective function is preferable because it is more general.

Surprisingly, a good objective function definition is not necessarily obvious. This can be problematic because the selection operators, described shortly, use the numeric score provided by the objective function to make decisions. Some circuits have very complex behaviors, which are not easily captured in a simple, closed form mathematical expression. Nonlinear behaviors such as hysteresis are a good example. Even simple combinational logic circuits are not immune. Consider evolving a digital circuit to implement an exclusive-or function. An initial choice for an objective function might be to apply all four input conditions (00, 01, 10, and 11) and record the number of correct responses. But what if one evolved circuit got the wrong response for the inputs 00 and 01, while another evolved circuit got the wrong response for the inputs 00 and 11. Which circuit is better? The objective function doesn't really provide much help.

Unfortunately, in some problems maximizing a desirable property may also maximize an undesirable property. For instance, circuits that operate at high speed also tend to have high power consumption. Obviously a high value should be awarded for correct operation at a high speed, but if low power consumption is also necessary, somehow this information should appear in the objective function's format. Put another way, if solution s and s' both operate correctly and at the same speed, but s consumes less power, then we want $\mathcal{O}(s) > \mathcal{O}(s')$. One way to accomplish this is to incorporate a penalty into the objective function. That is,

$$\mathcal{O}(s) = \text{fitness}(s) - \text{penalty} \quad (2.6)$$

where the penalty artificially decreases the natural fitness of the solution s. The penalty can be a function of the power consumed or it can be a fixed value large enough so the solution can't ever be selected. Another option is to tie the penalty to the generation number—i.e., high power consumption may impose a smaller penalty at the beginning of the search than it does near the end of the search. The exact form of the penalty is problem dependent.

Sewall Wright (1932) introduced the concept of a *fitness landscape* to capture the underlying dynamics of searching a solution space. This landscape formulates the solution set an abstract space, where two solutions are juxtaposed if they differ by only a single mutation. For example, if all solutions are encoded as binary strings, then two solutions are adjacent if they differ in one and only one bit position—i.e., their Hamming distance equals one. More formally, a fitness landscape consists of

- a large (albeit finite) set of solutions X;
- a fitness function $f : X \to \Re^+$ (the positive real number line); and
- the concept of a neighborhood between solutions.

Figure 2.5 shows an example 3-D landscape. Here x and y represent problem parameters so each point in the xy-plane is a problem solution. Adding a third dimension to show the fitness of each solution forms a fitness landscape. Of course most real-world problems have more than two parameters so the fitness landscape cannot be visualized. Nevertheless, the figure does show the complex structure that can exist in a fitness landscape, which illustrates how difficult it can be to find a globally optimal solution.

Landscapes are naturally formulated as metric spaces since the inherent distance metrics portrays our concept of a landscape with peaks, valleys, plateaus,

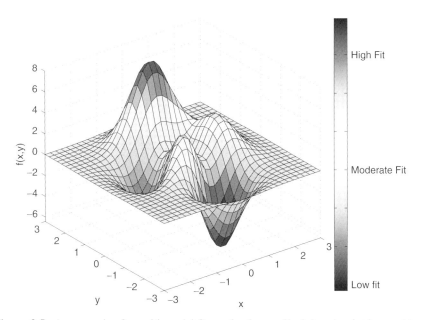

Figure 2.5. An example of a multi-modal fitness landscape. Each (x, y) point is a problem solution and the elevation of the landscape at that point indicates the corresponding solution's fitness. Highly fit solutions, which are high quality solutions, are at the higher elevations.

ridgelines and saddle points. This metric space formulation provides a framework for discussing optimization that matches our intuition. For instance, neighborhoods are defined in terms of objects being within close proximity to each other.

The technical literature often uses *move operator* instead of variation operator when discussing fitness landscapes. That is, searching in a fitness landscape is done with a move operator that transforms the current solution $x \in X$ into another candidate solution $y \in X$. A move operator is efficient if it can locate the optimal solution with a small amount of computational effort.

The best way of designing an efficient move operator is to have *a priori* knowledge of the fitness landscape's structure. If the landscape is "rugged", then it has lots of local optima and any move operator should make small steps to prevent missing good solutions. On the other hand, a "smooth" landscape has little fitness variation between neighboring solutions; large steps can be taken with little risk of missing good solutions.

A number of landscape characterization methods have been proposed and many of them use a metric space model. Weinberger [12] proposed a method that uses a random walk to gather statistical information. Starting at some randomly chosen solution x_t the walk next visits a randomly chosen neighbor. Repeating this process yields a sequence of fitness values $f_t f_{t+1} f_{t+2} \cdots$. Weinberger assumed that since there is some underlying distribution of fitness values, a random walk in any direction is sufficient to gather statistics. The degree of correlation between two genotypes z steps apart in this random walk is given by the correlation function

$$R(z) = \frac{\langle f_t f_{t+z} \rangle - \langle f_t \rangle^2}{\sigma_f^2}, \qquad (2.7)$$

where $\langle \cdot \rangle$ means the expected value over all pairs z steps apart. Highly uncorrelated landscapes are rugged whereas highly correlated landscapes are smooth. Other researchers believe a graph-theoretical perspective is the proper way to think about solution spaces [9]. However, for even moderate size optimization problems these graph-theoretical approaches are completely useless because the graphs they create are so large and so dense it is impossible to extract any meaningful information from them [7].

2.2.4 Selection

The variation operators depend on the individual's representation. Conversely, selection operators often depend on an individual's fitness. The most common selection methods are

- **Uniform Selection**

    This selection method is frequently used with evolution strategies. Here an individual is selected from the current population with equal probability. This is the only selection method that does not use fitness as a selection criteria.

- **Fitness Proportional Selection**

    The probability of selecting an individual is directly proportional to that individual's absolute fitness—i.e., the higher the fitness, the higher the probability of selection. The selection probability is easily calculated. Let f_i be the fitness of individual i. Then the probability of selecting individual i from a population of size N is

    $$\text{prob}(i) = \frac{f_i}{\sum_{j=1}^{N} f_j}. \tag{2.8}$$

    This selection method requires all fitness values to be strictly positive (to prevent a zero denominator). Fitness scaling can be used to handle cases where $f_j \leq 0$.

- **Fitness Ranking Selection**

    Individuals are sorted by absolute fitness value with the least fit individual in a population of size N given rank 1, the second least fit individual rank 2 and so on until the highest fit individual has rank N. Individuals are selected with a probability tied to their rank index instead of to their absolute fitness value (as is done in fitness proportional selection). There are both linear and nonlinear formulas for calculating the selection probability (see [1], p. 188). A nonlinear example is exponential ranking where the probability of selecting the individual i with rank r is

    $$\text{prob}(i) = \frac{1 - \exp(-r)}{C}, \tag{2.9}$$

    where C is a normalization constant to make sure $\sum_r \text{prob}(i) = 1.0$.
    A different type of ranking is used with multiobjective optimization problems (MOPs). Every solution of this type of problem has multiple attributes such as speed, gain, power consumption and selling price. The objective is to search for those solutions that have the best possible value for every attribute. (In some cases an optimal attribute value is a maximum while in other attributes an optimal value is a minimum.)
    Let X be the set of all possible solutions to a MOP. Each $x \in X$ has an associated attribute set $A = \{a_1, a_2, \ldots, a_n\}$. A solution x is said to *dominate* a solution x' (denoted by $x \succ x'$) if every a_k is better than or equal to every a_k' but there exists at least one m such that a_m is strictly better than a_m'. (a_k is only compared against a_k' $\forall k$.) The best solutions are non-dominated. If $x \not\succ x'$ and $x' \not\succ x$, then x is said to be "indifferent to" x' (denoted by $x \approx x'$). By definition, non-dominated solutions are indifferent to each other. The set of globally non-dominated solutions are referred to as the *Pareto optimal set*. More formally, the Pareto optimal set $\mathcal{P}^*$ is defined as

    $$\mathcal{P}^* = \{x \in X | \neg \exists x' \in X \text{ where } x' \succ x\}.$$

Ranking based on dominance is done using Goldberg's technique [4]. A pairwise comparison identifies all the non-dominated solutions in the current population. All non-dominated solutions are assigned rank 1 (the highest rank) and then removed from further consideration. A pairwise comparison amongst the remaining solutions will find a new set of non-dominated solutions, which are assigned rank 2 and then no longer considered. This assignment process repeats until all individuals have a rank. The rank-based probability formulas mentioned above cannot be used because more than one individual can have the same rank.

- **Truncation Selection**

    This deterministic selection method selects individuals from the previous population to construct the next population. The μ parents from $P(t)$ are subjected to variation operators that produce λ offspring. All $\mu + \lambda$ are collected together and sorted by fitness values. The μ best-fit individuals are kept to form $P(t+1)$ and the rest are discarded.

- **Tournament Selection**

    Tournament selection takes a random uniform sample of $q > 1$ individuals from the population and then selects the highest fit individual. Binary tournament selection ($q = 2$) is widely used although in principle any q value is acceptable. Large q values increase the chances of having above-average fitness individuals in the sample set.

With the exception of truncation selection, all of the selection methods are stochastic, which means the globally optimum solution, if found, could be accidentally lost. One way to prevent this is by using *elitism* where a small number of the best fit individuals are copied unchanged from $P(t)$ to $P(t+1)$. (Notice that truncation selection is naturally elitist.) This elitism concept can be carried even further to form a *steady-state EA* where only a fraction of the worst fit individuals are replaced each generation. The percentage of the population that does get replaced is called the *generation gap*.

A *selection pressure* characterizes every selection method. Selection pressure refers to how much "weight" is given to highly fit individuals. High selection pressure means individuals with a high (relative) fitness have a high selection probability. It is important that the selection pressure not be too high or the EA could prematurely converge. This issue will be discussed in greater detail later in the chapter.

2.2.5 Population

The population is a set of solutions that evolves as the EA runs. Actually the term *multiset* is more appropriate because the population may contain more than one copy of the same individual. The initial population is randomly generated to maximize the diversity. Diverse populations have many different solutions that, at least initially, produce a more thorough search. However, care must be taken with constrained optimization problems because a completely random population might contain illegal solutions. It is easy to create a legal, albeit random, population with N individuals. Start with one single legal solution—if necessary, constructed by hand—and duplicate it $N-1$ times. Since this is a constrained optimization problem, the variation operator is purposely designed to create only legal offspring from legal parents. Now apply this variation operator to each individual a random number of times to create a random population composed entirely of strictly legal individuals.

The population size is usually fixed in EHW problems. A consequence of sampling a finite population without some form of mutation is, over time, the population will eventually contain only one type of individual. This loss of genetic diversity is called *genetic drift*, and the rate of loss is inversely proportional to the population size; smaller populations succumb to genetic drift effects more rapidly.

2.2.6 Termination Criteria

Three termination criteria commonly used are

1. **The Algorithm has Converged**

    Convergence is assumed to exist if there is no improvement in the search over the previous k generations. ($k = 10$ is frequently used.)

2. **A Fixed Number of Generations have been Processed**

    The EA in this case runs for a predetermined number of generations. In EHW problems this number typically ranges from around 30 generations up to several thousand generations depending on the complexity of the circuit undergoing evolution. There are no formulas to compute this number because there are tradeoffs involved: too small a number may provide enough time to search for a good solution, while too large a number wastes computing time if the algorithm quickly converges.

3. **A Sufficiently Good Enough Solution has been Found**

    This solution may not necessarily be optimal, but it is good enough to satisfy the designer's requirements.

2.3 GETTING THE EA TO WORK

There are a number of factors to consider when formulating an EA to solve an EHW problem. There are many options available, which means trade-offs are necessary. For instance, a small population size will reduce the EA running time, but the resulting search may not be very thorough. A high selection pressure can focus the search, but if it is too high the EA could prematurely converge.

EAs have a number of tunable parameters (population size, mutation probability, etc.) that must be determined before any real search operations commence. Unfortunately there are no canned answers, no design formulas, and few rules available—EA design is hardly a pedantic process! The situation is not, however, completely hopeless because some heuristics do exist.

The first efforts should concentrate on a suitable representation. With commercial devices such as field-programmable gate arrays there is no real choice—the vendor has already predefined the data structure. But other cases the designer has complete freedom to choose a data structure. For example, suppose we want to design an amplifier circuit using MOSFETs. The designer is not constrained to use any particular data structure because the circuit is constructed from discrete components. It may be necessary to convert the data structure to another format to make it compatible with a circuit simulator, but that has no impact on the EA because the EA manipulates genotypes. The issue here is to make sure whatever representation is chosen does completely describe a circuit configuration. Hybrid representations will most surely be needed if the EA investigates both component values and different circuit configurations. Real number representations are not always necessary. Indeed, integer only representations are fine—and the EA may even run faster—if only component values are selected.

The variation operators are, of course, linked to the representation chosen, but it is in this area some work may have big dividends. A substantial amount of empirical evidence suggests some form of recombination is beneficial, but a moderate amount of mutation should always be used to keep the search from stagnating. This is particularly true when the EHW is used in any fault tolerant system [6]. Many different recombination forms have been proposed, but the designer should explore novel forms to see how well they work. Panmictic forms of recombination should not be ignored. Most EAs use more than one variation operator, each applied with some probability. Don't be afraid to experiment with both the type of operator and its probability of use. One effective search strategy is to use an operator that makes large moves in the solution space as the primary search operator and combine this with a local search operator that only makes small moves.

It is also wise to consider using some form of variation operator adaptation—e.g., as was done in Eq. (2.3). Remember the initial population is (ideally) uniformly distributed throughout the solution space. It makes sense then to have large movements in the solution space. However, near the end of the search the population should be located near very good solutions (which means only small moves should be made to prevent missing interesting solutions). There is some advantage in being able to tailor the movement step sizes at different times during the search.

Arguably the most difficult thing to achieve is a proper balance between *exploration*, where new individuals are created for evaluation, and *exploitation*, where the search concentrates on those regions of the solution space where good solutions are known to exist. Both population size and selection pressure affect this balance. Population size is often dictated by a desired run time—large populations have large run times. There may not be much flexibility in choosing the population size. Selection pressure can be influenced. A very strong selection pressure, such as found in roulette-wheel selection, should be avoided because it quickly loses population diversity which causes the EA to converge too rapidly. Some form of tournament selection would be a reasonable first choice to investigate if there is little fitness variance in the population. Conversely, with large fitness variance a ranking selection scheme should be used. Studies indicate exponential ranking is very good at maintaining some population fitness diversity [2].

The many components of an EA are interrelated; changing one parameter inevitably affects some other parameter. Domain experts are often needed to assist in fine-tuning the EA—that is, an application expert and an evolutionary algorithm expert must work together to design an efficacious EA. The application expert can identify the key operational characteristics needed for the evolved circuit. In fact, their help with defining an appropriate objective function will undoubtedly be their most significant contribution. Lacking an application expert, the EA designer must resort to a literature search to see what was done to solve similar EHW problems.

2.4 WHICH EA IS BEST?

The term "evolutionary algorithm" is actually a generic term. Most EAs in use today descend from three independently developed approaches: *evolutionary programming*, *genetic algorithms*, and *evolution strategies*. In the past the distinction between these algorithms was pretty clear—genetic algorithms used roulette-wheel selection and the primary reproduction operator was recombination; evolutionary programming only used tournament selection, and so on. Now the distinctions are becoming somewhat blurred and researchers are not so adamant about sticking with one particular type of algorithm.

A slightly different type of EA that has been successfully used with EHW problems is *genetic programming* [8]. This EA is also based on neo-Darwinian principles, but its real purpose is to evolve a computer program that solves the problem at hand. The user provides five inputs: (1) a set of terminals (zero-argument functions, initial variable values, etc.), (2) a set of primitive functions for each branch of the evolving program, (3) the objective function, (4) run control parameters, and (5) the termination criteria. The initial population is a set of randomly generated computer programs that are usually represented as rooted, point-labeled trees with ordered branches. Each tree is mapped to a labeled cyclic graph that represents a physical circuit. Reproduction is performed with random mutation and crossover operators.

The execution time for a genetic programming algorithm used to solve an EHW problem can last for days—even though the algorithm is usually running on a large multiprocessor system. One contributing factor is genetic programming populations are typically two or three orders of magnitude larger than the populations used with other types of EAs. Nevertheless, genetic programming has evolved circuitry competitive with what humans have designed.

Genetic algorithms are arguably the most prevalent EA used by EHW practitioners. Does this mean genetic algorithms are the best EA to use? Unfortunately, that question is not easy to answer. The no-free-lunch (NFL) theorem states there is no search algorithm that works best for *all* optimization problems. More specifically, the NFL theorem says if an algorithm does particularly well on average for one class of problems, then it must do worse on average over the remaining problems [10]. This means debating which algorithm is best for solving all EHW problems is a complete waste of time. Besides, any performance comparison between search algorithms is meaningless unless certain precautions are taken [5]. A designer's time would be far more wisely spent on fine-tuning an EA to educe its best possible performance rather than worrying about which particular type of EA to use.

REFERENCES

1. Bäck T, Fogel D and Michalewicz T (Eds.) 2000. *Evolutionary Computation 1*, Institute of Physics
2. Blickle T and Thiele L 1995. "A Mathematical Analysis of Tournament Selection", *Proceedings 6th International Conference on Genetic Algorithms*, San Mateo, CA: Morgan Kaufmann, 9–16
3. Eiben A and Smith J 2003. *Introduction to Evolutionary Computing*, Springer-Verlag, Berlin
4. Goldberg D 1989. *Genetic Algorithms in Search, Optimization and Machine Learning*, Reading, MA: Addison-Wesley
5. Greenwood G 1997. "So Many Algorithms, So Little Time", *ACM Software Engineering Notes* 22(2), 92–93
6. Greenwood G, Ramsden E, and Ahmed S 2003. "An Empirical Comparison of Evolutionary Algorithms for Evolvable Hardware with Maximum Time-to-Reconfigure Requirements", *Proceedings 2003 NASA/DOD Conference on Evolvable Hardware*, 59–66
7. Greenwood G 2005, "On the Usefulness of Accessibility Graphs With Combinatorial Optimization Problems", *Journal of Interdisciplinary Mathematics* 8(2), 277–286
8. Koza J, Keane M and Streeter M 2003, "What's AI done for me lately? Genetic programming's human-competitive results", *IEEE Intelligent Systems* 18(3), 25–31
9. Stadler B and Stadler P 2002, "Generalized Topological Spaces in Evolutionary Theory and Combinatorial Chemistry", *Journal of Chemical Information and Computer Sciences* 42, 577–585
10. Wolpert D and Macready W 1997. "No Free Lunch Theorem for Optimization", *IEEE Transactions on Evolutionary Computation* 1(1), 67–82

11. Wright S 1932, "The Roles of Mutation, Inbreeding, Crossbreeding and Selection in Evolution", *Proceedings of the 6th International Congress on Genetics*, 356–366
12. Weinberg E 1990, "Correlated and Uncorrelated Fitness Landscapes and How to Tell the Difference", *Biological Cybernetics* 63, 325–336
13. Bäck T 1998, "An Overview of Parameter Control Methods by Self-Adaptation in Evolutionary Algorithms", *Fundamenta Informaticae* 35, 1–15

CHAPTER 3

RECONFIGURABLE DIGITAL DEVICES

Aims: *Reconfigurable devices are the second key component of EHW. This chapter covers the internal architecture, capabilities and programming requirements of digital devices. (Analog devices are discussed in the next chapter.) The chapter begins with a survey of reprogrammable devices from major IC manufacturers and concludes with a description of the POEtic chip fabricated as part of a European research project.*

3.1 BASIC ARCHITECTURES

Before talking about specific devices that might be used to create evolvable hardware, it is useful just to remind ourselves of some of the basics of logic design and how technology has got to its current position.

In the past, engineers designed digital electronic circuits using various top down decomposition methods that resulted in a small city of logic chips. At that time the 74 series and the 4000 series standard digital gates were used for random glue logic (random looking logic structure, usually used to interface different parts of a system). But for commercial systems the requirement was for an *application specific integrated circuit* (ASIC) that could be easily manufactured to perform a specific function. Unfortunately the manufacturing time for the design and implementation of a complete chip was (and still is) prohibitively expensive for

---

Introduction to Evolvable Hardware: A Practical Guide for Designing Self-Adaptive Systems, by Garrison W. Greenwood and Andrew M. Tyrrell
Copyright © 2007 Institute of Electrical and Electronics Engineers

anything except large-scale production runs. For small-scale production runs a cheaper device with a shorter design time is needed.

There are two types of programmable devices of interest: The *programmable logic device* (PLD) and the *field-programmable gate array* (FPGA).

A generic PLD structure can be seen in Figure 3.1. The inputs to the chip enter the wired AND gates to create a product term, and these terms are then OR'd together to give a sum-of-products term which can either be fed directly through to the output pin, or through the register (flip-flop) to the pin. Either way the output can then be fed back to the input by inserting the signal into the wired AND gates. While there are minor difference in the structure of different PLDs, the AND-OR structure is common. Why you may ask? Before going on to describe the developments of PLDs to FPGAs let's consider some basic logic design. (NOTE: if you are familiar with logic design or are not interested in the details you can skip the rest of this section).

Any Boolean function can be defined by a truth table. A unique algebraic expression for any Boolean function can be obtained from its truth table by using an OR operator to combine all minterms for which the function is equal to 1. Remember for an n-variable function there are 2^n rows in the truth table and 2^n minterms.

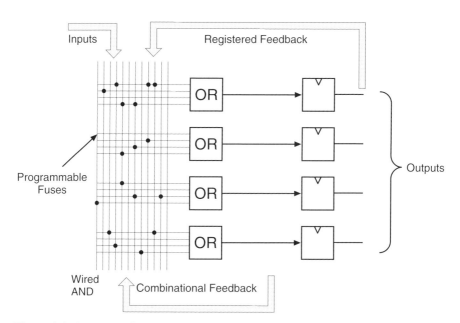

Figure 3.1. Structure of a Programmable Logic Device (PLD) The circuit inputs (and their inverts), the register outputs and the combinational logic outputs are all fed into the wire-AND array.

Example:

x	y	z	MINTERMS	NOTATION	F_1
0	0	0	$\bar{x}\,\bar{y}\,\bar{z}$	M_0	0
0	0	1	$\bar{x}\,\bar{y}\,z$	M_1	0
0	1	0	$\bar{x}\,y\,\bar{z}$	M_2	0
0	1	1	$\bar{x}\,y\,z$	M_3	1
1	0	0	$x\,\bar{y}\,\bar{z}$	M_4	0
1	0	1	$x\,\bar{y}\,z$	M_5	1
1	1	0	$x\,y\,\bar{z}$	M_6	1
1	1	1	$x\,y\,z$	M_7	1

$$F_1 = M_3 + M_5 + M_6 + M_7 = \bar{x}\,y\,z + x\,\bar{y}\,z + x\,y\,\bar{z}$$
$$\overline{F_2} = M_0 + M_1 + M_2 + M_4 = \bar{x}\,\bar{y}\,\bar{z} + \bar{x}\,\bar{y}\,z + \bar{x}\,y\,\bar{z} + x\,\bar{y}\,\bar{z}$$

This example demonstrates an important property of Boolean algebra:

ANY BOOLEAN FUNCTION CAN BE EXPRESSED AS A SUM OF ITS 1-MINTERMS

Any Boolean expression can be converted into this sum-of-minterms form by first expanding the given expressions into a sum of AND terms. Then each term that is missing one or more variables has to be expanded further, by ANDing the term with an expression such as $x + \bar{x}$, where x is one of the missing variables.

Example:

Express the Boolean function $F = x + y\,z$ as a sum of minterms.

Assume the function has three variables: x, y and z. The first term is missing two variables while the second term is missing one variable. Thus, add $(y + \bar{y})$ AND $(z + \bar{z})$ to the first term and $(x + \bar{x})$ to the second term.

$$F = x(y + \bar{y})(z + \bar{z}) + (x + \bar{x})y\,z$$
$$= x\,y\,z + x\,\bar{y}\,z + x\,y\,\bar{z} + x\,\bar{y}\,\bar{z} + x\,y\,z + \bar{x}\,y\,z$$
$$= \bar{x}\,y\,z + x\,\bar{y}\,\bar{z} + x\,\bar{y}\,z + x\,y\,\bar{z} + x\,y\,z + x\,y\,z$$

This is now in canonical form. The same (similar) procedure can be carried out on the maxterms, in this case we end up with the product-of-maxterms—again the canonical form. It is the sum-of-products form that is of most interest to us here. This structure has AND gates, which implement the minterms, feeding an OR gate. This is precisely the structure depicted in Figure 3.1.

3.1.1 Programmable Logic Devices

We have seen that programmable logic devices contain gates, and in some cases flip-flops, arranged so that the interconnections between the components can be altered to implement various logic functions.

A general requirement of a PLD is to have a means of changing the interconnections to form a different logic configuration. The original permanent but selectable way of achieving this was to manufacture the devices with semiconductor fuses. Initially these fuses are intact within the device, which provides extensive interconnections. Selected fuses are then "blown" by the user to obtain the desired interconnections using special PLD programmer units (usually PC based). Once blown, however, the connection cannot be remade.

This type of device became very useful for random glue logic, since it can be simply (and permanently) programmed in a few seconds by burning the specific fuses required. The only problem with the simpler PLDs is the output architecture is fixed—that is, outputs are pre-defined as combinational outputs or registered outputs. The available mix of outputs may not be exactly what the designer needs. For example, the popular PAL16R4 has eight possible outputs, four combinational and the other four registered. The PAL16R6 also has eight possible outputs but six are registered and two are combinational. But, what if a designer really needs five registered outputs and 2 combinational outputs? Unfortunately no such PLD exists.

Enter the *complex programmable logic device* (CPLD). The CPLD is an extension of the PLD but it provides greater flexibility. One example of the CPLD is shown in Figure 3.2. At first glance one would think that the logic was contained

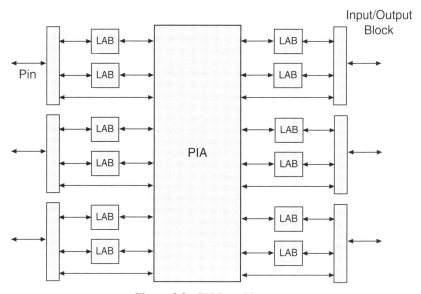

Figure 3.2. CPLD architecture.

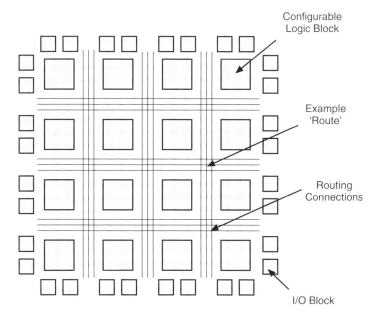

Figure 3.3. Basic FPGA architecture.

in the center, in fact the logic for the CPLD is contained in the Logic Array Blocks (LABs), and the center is the Programmable Interconnect Array (PIA) where the inputs and output to the logic blocks are connected. Each LAB is very similar in construction to the simple PLD. One primary advantage a CPLD has over the PLD is the CPLD lets the designer program the outputs individually. That means a designer can have a device with five registered outputs and two combinational outputs if that is what is really needed.

3.1.2 Field Programmable Gate Array

The FPGA[1] is organized as an array of logic blocks as shown in Figure 3.3. Programming an FPGA requires programming three things: (1) the functions implemented in logic blocks, (2) the signal routing between logic blocks, and (3) the characteristics of the input/output blocks (e.g., a tri-stateable output or a latched input).

Some devices use a fuse technology, which means they are *one time programmable* (OTP). Another FPGA programming method, and the one used in almost all devices now, is by downloading the configuration information into static RAM (SRAM). The main advantage of an SRAM implementation is the

---

[1] Most of the diagrams in this section are taken from Xilinx datasheets. See http://www.xilinx.com for more information.

FPGA design now becomes reprogrammable. Typically the entire FPGA is reprogrammed, although some devices can be partially reconfigured. Partial reconfiguration means the designer can choose to reprogram only a portion of the device rather than the entire device. This capability can drastically reduce the reprogramming time in larger FPGAs. However, not all FPGAs have this capability.

Recently Lattice Semiconductor Inc. introduced the LatticeXP FPGA family that combines FLASH memory and SRAM. Normal operation is executed out of SRAM, but the entire SRAM can be reprogrammed in a about one millisecond from the internal FLASH. One nice feature is the FLASH can be reprogrammed while the device is online and executing out of SRAM[2].

Configurable Logic Blocks

Figure 3.3 showed a matrix of logic components connected through random routing to form a digital logic circuit. The logic components depicted could be simple logic gates which are unchangeable, for example, the NAND gate could be used throughout the whole array. The functionality of the circuit would then be described through the routing rather than through the logic. Unfortunately to implement such a fine grained FPGA would require more routing which in turn requires more routing matrices. It is better to use more complex logic components, reducing the routing requirements and therefore its complexity.

These logic components are known as configurable logic blocks (CLBs). The configurable logic block is the smallest unit of programmable logic found on the FPGA. There are two main methods of making the logic block 'configurable', through the use of static Look-Up Tables (LUTs) or through the use of multiplexers. Figure 3.4, shows a Xilinx XC4000 [2] CLB, the three 'logic function' blocks are the LUTs, these are more commonly known as the F-LUT, the G-LUT and the H-LUT.

The F and G LUTs are connected through the routing to other CLBs in the FPGA through the F1–F4 and G1–G4 inputs. There are further inputs to the CLBs (C1-C4). These are the control inputs which control register enable (EC), set/reset (SR), a serial data input (DIN) and another configurable input (H0). The LUTs and multiplexers together form a highly flexible logic block.

The Algotronix CAL device, which was bought out by Xilinx and re-named the XC6200 uses a slightly different method of configuration. Figure 3.5 shows the schematic of the logic contained within the 6200 CLB. Here the logic is formed using multiplexers rather than LUTs. In the 6200 CLB there are only three inputs and one output from the CLB, (not including the clock and clear lines).

Finally, consider the Xilinx Virtex II device. These configurable logic blocks (CLBs) are organized in an array and are used to build both combinatorial and synchronous logic designs. Each CLB element is tied to a switch matrix to access the general routing matrix, as shown in Figure 3.6. A CLB element comprises 4

---

[2]Unfortunately, at the present time these devices have limited use in EHW applications because the number of FLASH reprogramming cycles is rather low.

BASIC ARCHITECTURES   41

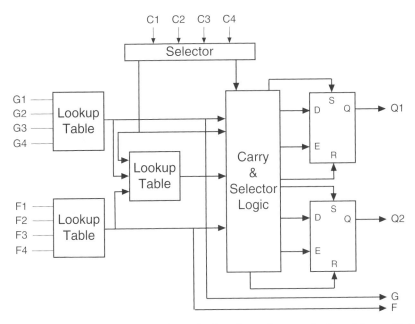

Figure 3.4. Xilinx XC4000 CLB schematic. Details on the carry logic and the multiplexers has been omitted for clarity.

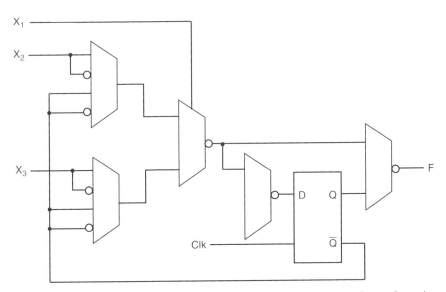

Figure 3.5. XC6200 Function Unit, multiplexers provide the elements of reconfiguration.

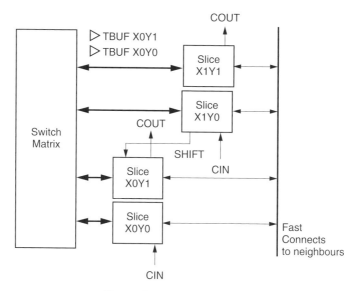

Figure 3.6. Basic Virtex CLB.

similar slices, with fast local feedback within the CLB. The four slices are split in two columns of two slices with two independent carry logic chains and one common shift chain.

Slice Description

Each slice includes two 4-input function generators, carry logic, arithmetic logic gates, wide function multiplexers and two storage elements. As shown in Figure 3.7, each 4-input function generator is programmable as a 4-input LUT, 16 bits of distributed SelectRAM memory, or a 16-bit variable-tap shift register element.

The output from the function generator in each slice drives both the slice output and the D input of the storage element.

In the context of what we are talking about here, the structure of a particular CLB is not important, the main thing to note is that they provide a flexible way, through a programmable interface, of creating a logic function.

The other important aspect of a system is the connectivity of these functional blocks to form a complete system.

Routing Architectures

The main method of routing in an FPGA is to use metallization layers on the silicon to create hundreds of uncommitted wires. This creates the array of wiring which can then be used to convey information around the FPGA. But there is another way, the Algotronix Cal device, or XC6200 used a very different method

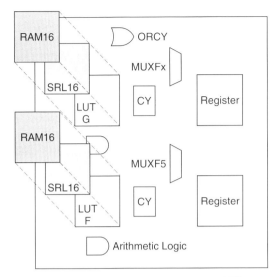

Figure 3.7. Functional block structure of Virtex.

of routing. The regular method of routing will be examined and compared with the XC6200 routing method.

Regular Routing

The regular routing system is shown in Figure 3.8 is the routing scheme of the Virtex device (it is only shown in the east-west direction) and is similar to many other current devices. The wires are divided into two different types: single lines and hex lines.

The single lines terminate at the adjacent CLB whereas the hex lines terminate at a CLB six places away in a given direction. In the Virtex there are actually 24 single lines in each compass direction and 12 hex lines. Inside the CLB there is a local routing box which routes communications from the global routing box to the local connections of the CLB. Here the single and hex lines can be connected to the inputs of the CLB.

This method of routing is very flexible because of the number of wires and the fact that they are bi-directional. The main implication this has on the design of digital electronic circuits is that the CLB logic can now be made at a slightly coarser grain because the routing is now available to fully utilize the logic cell. This means that FPGA designs will be more efficient than previously possible.

XC6200 Routing

The XC6200 device has a very different scheme for routing. In fact, the routing is completely hierarchical as shown in Figure 3.9. The routing at the lowest level is neighborhood routing (which, in total forms the largest number of routing connections). Cells which are on the 4×4 cell boundaries are also able to drive

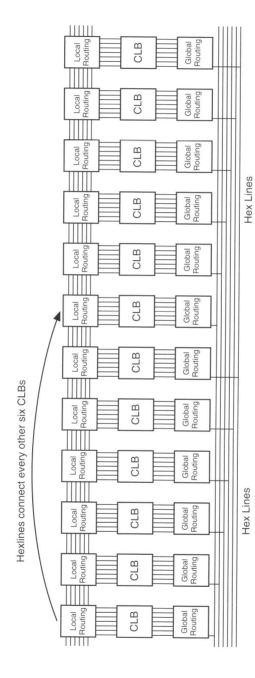

Figure 3.8. Virtex routing architecture is a typical sea of gates architecture, where the logic is surrounded in routing. There are two levels of routing, local routing (*single lines*) or global routing (*hex lines*) which terminate six CLBs away.

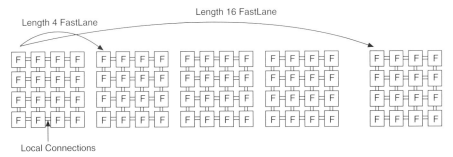

Figure 3.9. The XC6200 Hierarchical routing system. The XC6200 has different levels of routing organized in a hierarchical fashion. For example, there are local routes which run directly between each node and its neighbor and length 4 FastLANEs skip the four next CLBs. There are also Chip Level (CL) lines which run the whole width/height of the device.

their outputs onto length 4 FastLANEsTM, which connect the current 4×4 blocks to the surrounding 4×4 blocks. Cells on the 16×16 cell boundaries can similarly drive their outputs onto the length 16 FastLANEsTM.

This method of routing is interesting because all of the wires are unidirectional, meaning that there can be no contention within the device and that unlike a channeled architecture used on the Virtex the configuration will never destroy itself (electrically that is). Of course the routing is also limited, since there are only five outputs from each CLB (north, south, west, east and the magic output). This means that to increase the utilization of the logic it has been necessary to reduce the complexity of the logic held inside the CLB as in Figure 3.5.

Once again, in the context of what we are talking about here, the structure of a particular interconnect on a particular device is not important. The main thing to note is that the programmable interface provide a very flexible way of creating a system.

3.2 USING RECONFIGURABLE HARDWARE

We have talked so far about some generic digital design concepts and some specific devices that have been produced over the last few years which we might make use of to produce reconfigurable systems. The best way to bring these elements together is to give a simple, but real, example of how we might use one of these devices in such an application. The example we will consider now is a simple image processing application which evolves a certain functionality, in this particular case an image filter. This digital filter will be intrinsically evolved.

Evolutionary computation is able to improve or even replace human design of combinational circuits. However, this is usually achieved either through a computational effort that involves the sampling of a large number of individuals and the evolution for a great number of generations (as described in Chapter 2), or by devising new evolutionary techniques. For the latter, a sensible combination of the reconfigurable primitives and the evolutionary operators is essential to the success of intrinsic EHW design. The work in this example employs intrinsic EHW by devising an array of compact processing elements and an external genetic reconfiguration unit, which outperforms human design in terms of computational effort and implementation cost.

We approached the proposed problems using cartesian genetic programming (CGP), a concept originally introduced by Miller and Thomson [1]. Unlike conventional genetic programming where genotypes are represented as undirected tree graphs, in CGP genotypes are represented as rectangular array digraphs.

In practice the genotype just contains a list of vertex connections and functions. One instance of a function with, say, 3 inputs is called a *molecule*. A digital circuit can be encoded as an array of molecules with n rows and m columns where each molecule's input can only connect to molecules from columns to its left. More specifically, the molecules in the left most column can only receive data from the inputs to the system. Molecules in the other columns can received data from molecules in the p proceeding columns. (p is referred to as the "levels back" parameter.) No connections within the same column are allowed. These restrictions ensure the feed-forward structure typically found in pure combinational logic circuits and thus guarantee only combinational circuits will be evolved.

A digital circuit is encoded as $n \times m$ molecules, with data width w, and level back parameter p. This is abbreviated as "the setting for a gene is $n \times m @ w - p$". (In the example that follows one gene encodes one circuit).

A sample circuit with gene setting $2 \times 2@8 - 1$ is shown in Figure 3.10. This circuit, which is composed of a maximum of four molecules, has four inputs and

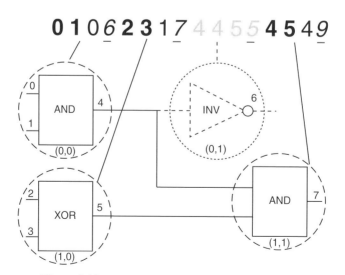

Figure 3.10. A sample digital circuit based on CGP.

one output. The underlined integers in italics represent functions: INV, AND, XOR and ADD which are functions 5, 6, 7 and 9 respectively. The other three integers in each group encode the inputs to this molecule. This circuit has a maximum of six inputs, connecting to the two molecules in column 1 (the left two), although in this sample circuit only four inputs are utilized. The output is molecule 7, so the INV gate (circled by a dotted line) encoded by the four grey digits is not used in this particular circuit. Bold integers are used inputs to the molecules and the remaining digits are ignored.

Sekanina [3] showed CGP was not only useful in evolving gate level designs, such as three-bit adders, but also for evolving functional level designs such as digital image filters. But the standard CGP architecture does require some modifications before it can be effectively used. For example, consider an image filter with nine 8-bit inputs (for the neighborhoods) and one 8-bit output. It is nearly impossible to provide the required 72 inputs and 8 outputs needed by a conventional CGP architecture. Consequently we used an extended CGP architecture that was modified for our own purposes. As shown in Figure 3.11, the processing elements (PEs) are indexed from the top left (index 9). The two inputs of every PE can be connected to one of the outputs from the previous l columns (if the level-back parameter is 2).

Since our goal is an intrinsic EHW processor design, we have chosen a convenient application of evolving image filters for certain types of noise. Gray scale images of 256×256 pixels (8 bits/pixel) are considered in this example. As with the conventional approach, a square neighborhood of 3×3, centered at a target pixel, is defined and therefore applied at X and Y. Every PE executes one function from Table 3.1.

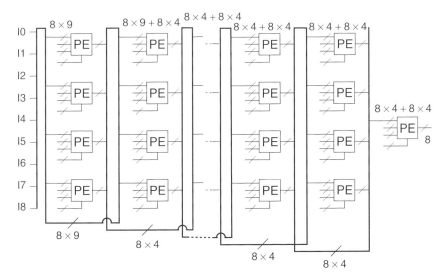

Figure 3.11. The Cartesian Genetic Programming Reconfigurable Architecture.

TABLE 3.1. Function Codes

Code	Function	Code	Function
F0: 0000	X >> 1	F8: 1000	(X + Y + 1) >> 1
F1: 0001	X >> 2	F9: 1001	X & 0x0F
F2: 0010	X	F10: 1010	X & 0xF0
F3: 0011	X & Y	F11: 1011	X	0x0F
F4: 0100	X	Y	F12: 1100	X	0x F0
F5: 0101	X ∧ Y	F13: 1101	(X & 0x0F)	(Y & 0xF0)
F6: 0110	X + Y	F14: 1110	(X & 0x0F) ∧ (Y & 0xF0)
F7: 0111	(X + Y) >> 1	F15: 1111	(X & 0x0F) & (Y & 0xF0)

```
 2 7 3    8 6 3    3 4 1    1 7 2    0 5 3    12 1 1    13 4 2    10 1 1 1   9 2 3    3 1 3 1    13 1 8 1
16 15 3   11 17 2   - - -    23 17 4  21 23 2  24 23 1   25 27 2   22 26 1   29 32 4
```

Figure 3.12. Chromosome encoded in an integer string.

Chromosomes are encoded into linear integer strings in triplets, such as shown in Figure 3.12, which is a description of a 4 × 6 CGP architecture.

A steady-state GA is employed. An initial population of 16 is generated randomly. The individual with the highest fitness is chosen as the parent and its mutated versions provide the new population. Previous experiments suggest that an adaptive mutation rate works well for hardware evolution [4]. In the case of the intrinsic evolution of image filters, an adaptive mutation rate was employed and the number of genes to be mutated was decided accordingly by this adaptive description:

$$\text{num} = c + p \cdot N_{genes} \cdot (1 - norm\_fit),$$

where

$$norm\_fit = \frac{fitness}{255 \cdot (H - 2) \cdot (W - 2)}$$

and c and p are user-defined parameters. Here c is set to be 4 and p is set to be between $0 \Rightarrow 1$. N_{genes} is the number of genes in every individual. For image filter design, usually two fitness functions are used, of which one is PSNR (Picture Signal Noise Ratio) and the other is MDPP (Mean Difference Per Pixel).

$$fitness_{PSNR} = 10 \cdot \lg \sqrt{S/N}$$

$$fitness_{MDPP} = 255 \cdot (H - 2) \cdot (W - 2) - \sum_{i=1}^{254} \sum_{j=1}^{254} |orig(i, j) - crpt(i, j)|.$$

For the sake of the generality and adaptability of this work, both fitness functions have been employed in the software simulation, but only the MDPP fitness function was implemented in hardware evolution, as it is computationally easier for hardware calculation and implementation. The image size used was 256×256, and subsequently, only an area of 254×254 pixels could be chosen for 3×3 neighborhood processing.

3.2.1 Design Phase

The EHW CGP structure is implemented in a Xilinx Virtex XCV1000 FPGA. As Figure 3.11 shows, the genotype layer contains the genetic information for evolution, which consists essentially of Processing Elements (PEs). Every PE cell is made up of two input multiplexers, one Functional Block (FB) and necessary interconnections, Figure 3.13. The FB contains a compact, and possibly redundant representation of the functions, one of which is to be chosen as the active function for this PE cell.

The evolutionary circuit operates on nine 8-bit inputs and one 8-bit output. For each PE, the multiplexer inputs will be the outputs from the previous two columns. Accordingly, both cfg1 and cfg2 should not exceed the number of the multiplexer inputs. The cfg3 input should be the binary representation of the number of functions in store. The fewer the functions the faster the evolution should be. Experience has shown that only functions F5, F7, F8 and F13 were exploited in the final evolved best circuits. Further functions can be excluded but this is dependent on the resource requirements, as there is a trade-off between the functionality and the complexity of the hardware structure.

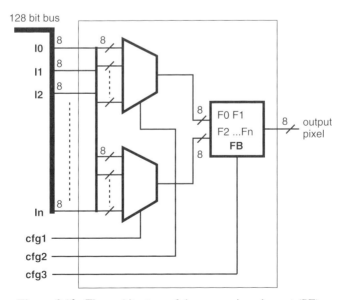

Figure 3.13. The architecture of the processing element (PE).

Synthesis Report

The main feature of this chip—that is, the 128-bit bus from the memory prefetch circuitry—really stands out. It supplies the PE array with a new 3×3 mask frame in every clock cycle. The entire kernel, therefore, is loaded in a single clock cycle and in each subsequent cycle a 3×3 block is brought in from the larger image until the last pixel and its neighborhood are loaded. Each PE is allotted eight bits of the 128-bit bus, so they are directly supplied with data individually. A 32-bit output bus for each column is another feature of this device. The output buses from the previous two columns are prepared for the configuration of the PEs. The functional configuration bus makes this structure compact as well, though it only takes 2 bits for each PE. $254 \times 254 + 1$ clock cycles are taken to download masks for evaluation from the whole image. The compact architecture uses 23 to 41 slices for each PE on the Xilinx Virtex FPGA XCV1000.

3.2.2 Execution Phase

The execution phase, namely the Genetic Processing Unit (GPU), is realized by the hardware implementation of a Genetic Algorithm. Due to pipelining, parallelization and no function call overhead, a hardware GA yields a significant speedup over a software implementation [5], which is especially useful for the real-time applications. While FPGAs are not as fast as typical ASICs, they still hold a great speed advantage over functions executing in software. In fact, a speedup of 1 to 2 orders of magnitude has been observed when frequently used software routines have been implemented on FPGAs.

GPU Components

The GPU system consists of seven modules, as shown in Figure 3.14:

> **Interface Memory (IM):** This is the central control unit. It responds to the start-up signal from the front end by asking the Input Buffer to download input data, and signals the front end to shut down when the evolution is accomplished. The initial population is produced by the Random Number Generator (RNG), read into and stored in the IM without being initially evaluated.
>
> **Input Buffer (IB):** The IB communicates with the front end at the request of the IM and reads in the original and distorted images, the mutation rate, and the RNG seed.
>
> **Random Number Generator (RNG):** The RNG reads the seed from the IB. After the seed is loaded, the RNG module employs linear cellular automata (CA) to generate a sequence of pseudorandom bit strings. The CA uses 10 cells that change their states according to the rule 90 and rule 150 by

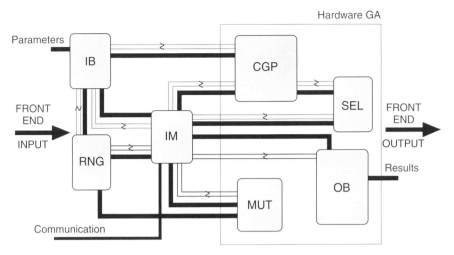

Figure 3.14. The GPU hardware architecture.

Wolfram [6]. It generates the initial chromosomes for the IM. At the request of the Mutation module (MUT), it sends out mutation points, the genes to be mutated and the possible mutated genes.

Mutation (MUT): For each chromosome, it reads and uses the parameters from the RNG and IM. The adaptive mutation rate is employed. Chromosomes would be sent back to the IM for further processing.

Evaluation (CGP): This is the core of EHW. It reads chromosomes from the IM and configures the CGP hardware structure. For each chromosome, or each potential circuit, the distorted bitmap is applied for evaluation.

Fitness Calculation (SEL): MDPP fitness function is used to calculate the MDPP. The best chromosome is sent back to the IM.

Output Buffer (OB): The evolution is shut down when a specific number of generations are reached. The OB will output the best chromosome and the filtered image.

The GPU Hardware Architecture

The GPU modules were designed to correlate well with the CGP operations, that is simple and easily scalable, and have interfaces that facilitate parallelization. They were also designed to operate concurrently, yielding a coarse-grained pipeline. The Interface Memory (IM) acts as the main control unit during start-up and shutdown and communicates with the front end. After start-up and shutdown, control is distributed. All modules operate autonomously and asynchronously. The basic functionality of the GPU architecture is as follows:

1. The front end signals the IM that the evolution is to start. The IM accepts the request, asks for the parameters from the front end and stores the parameters into the IB.
2. After loading the parameters, the IM notifies the CGP, the MUT and the RNG. Each of these modules requires its user specified parameters from the IB.
3. The initial population is randomly generated by the RNG and stored in the IM for evaluation and evolution.
4. The IM starts the pipeline by requesting the CGP to download the chromosomes and then the CGP passes them to the SEL to calculate and store their fitness.
5. The SEL selects the best individual and sends it back to the IM.
6. When the IM receives all the chromosomes, ready for the next generation, it signals the MUT to start mutation.
7. The genes to undergo mutation are sent to the MUT by the IM, and the MUT mutates them within a set range and returns them to the IM.
8. The CGP is working concurrently with the MUT.

This continues until the IM determines that the current GPU run is finished. It shuts down and signals the front end to read the final population back from the OB.

3.2.2.1 The Inter-Module Communications

The modules in Figure 3.14 communicate via a simple asynchronous handshaking protocol similar to asynchronous bus protocols used in many computer architectures. This protocol is shown in Figure 3.15.

The chromosome communication among modules is through one 256-bit bus, finished in only one clock cycle. Pixels go from one module to another through an 8-bit bus.

S module sends Data to R module

a. S raises a request signal to R
b. R agrees and raises an acknowledgement
c. S lowers its request
d. S transfers data to R
e. Data sent, R lowers the acknowledgement

Figure 3.15. The inter-module handshaking protocol.

Synthesis Report

When HGA starts up, the IB reads in the bitmap image within 256×256 clock cycles. The IM takes $16 \times 3 \times 4$ clock cycles to generate the initial population. While the CGP is evaluating the chromosomes (254×254 clock cycles), the MUT is executing reproduction ($16 \times 10 \times 7$ clock cycles). The entire GPU used 50.72% of the XCV1000 (6322 in slice), of which the CGP takes 1322 slices. This allows more complex CGP evolution and permits processing chromosomes in parallel.

3.3 EXPERIMENTAL RESULTS

We designed the experiments using additive noise distorted bitmaps. Gaussian noise and uniform noise are both additive noise, which is independent of the image itself. In many cases, additive noise is evenly distributed over the frequency domain, whereas an image contains mostly low frequency information. Hence, the noise is dominant for high frequencies and can be suppressed by low-pass filters. This can be done either with a frequency filter or with a spatial filter. Generally, a spatial filter is preferable in evolvable hardware applications, as it is computationally cheaper than a frequency filter. In both noise categories, we used the bitmap of Lena as the target image at different distortion levels for the generality of the EHW architecture. All results were compared with the filtered results from the conventional image filters, such as Gaussian, median and mean filters.

Figure 3.16(a) shows the original Lena bitmap. Figure 3.16(b) is the Lena bitmap distorted by Gaussian noise with mean 0 and $\sigma = 16$. Figure 3.16(c) is the resulting image filtered by the EHW filter. Figures 3.16(d), 3.16(e) and 3.16(f) show results from Median filter, Mean filter and Gaussian filter respectively. The MDPP is 15734299, 16040034, 16039677, 16017517 and 159980366 from Figures 3.16(b) to 3.16(f).

This example has presented a novel approach to combinational digital circuit design based on the technique of Intrinsic Evolvable Hardware. General-purpose image filters lack the flexibility and adaptability for un-modelled noise type. The EHW architecture evolves filters for certain types of noise without *a priori* information. In most industrial applications where design time and labor are of importance, this approach is preferable.

The EHW approach employs fewer hardware slices, thus is a computationally cheaper alternative. Generally, for certain types of noise, the evolved filters outperform conventional filters.

While this was a specific example (often the best type to illustrate points) there are a number of general points that we might point out.

- First and most important, it actually works! We have evolved a function, from scratch, actually on hardware (a Virtex FPGA) and its performance is at least comparable with human designs.

Figure 3.16. The Lena picture.

- It shows that we can create a reconfigurable system (evolvable systems after all are simply continuously reconfigurable systems) from commercial off-the-shelf (COTS) components.
- It is possible to embody a complete evolvable algorithm in hardware.
- Limitations are however required to execute such algorithms in hardware, such as the type of evolutionary algorithm chosen (e.g. in this case CGP), the resources that might be changed (e.g. the number of levels back that inputs might appear), the functions that might be produced (e.g. the possible function we let the CLBs produce) etc.
- For COTS devices it is difficult to have evolution working on the interconnect (e.g. in this example the configuration was fixed and only the function within CLBs were evolved) since in most cases random configuration streams might cause multiple outputs to be connected together with a resulting destruction of the device!

We have so far considered COTS devices for possible use in reconfigurable systems. We will see more examples of their use in Chapter 5. Let's now consider a purpose built device that has been developed for biologically-inspired systems, including reconfigurability. The device is called POEtic.

3.4 FUNCTIONAL OVERVIEW OF THE POETIC ARCHITECTURE

The implementation of bio-inspired systems in silicon is quite difficult due to the sheer number and complexity of the biological mechanisms involved. Conventional approaches exploit a very limited set of biologically plausible mechanisms to solve a given problem and often cannot be generalized because a design methodology for bio-inspired computing machines is lacking. This shortcoming is due to the heterogeneity of the hardware solutions adopted for bio-inspired systems, which is itself due to the lack of architectures capable of implementing a wide range of bio inspired mechanisms.

To partially overcome this problem, the goal of the "Reconfigurable POEtic Tissue" (or "POEtic" for short) [16], recently completed under the aegis of the European Community, was the development of a flexible computational substrate inspired by the evolutionary, developmental and learning phases in biological systems.

The overall architecture of the POEtic tissue, described shortly, was conceived after contemplating the basic features that might be required—or at least proved useful—in bio-inspired systems. These features can be summarized as follows.

- Reconfigurability
- Multi-cellular scalable structure.
- Possibility for implementing phylogenetic, ontogenetic and epigenetic mechanisms, separately or in any combination, these constitute the three main axes of biological organization.

- Layered hardware organization that resembles the three previous axes.
- Massive I/O interaction with the external environment.

The high-level organization of the POEtic tissue design is depicted in Figure 3.17. As can be seen from this figure, the tissue is composed of three main parts:

- **Environment Subsystem:** This is the component of the tissue that is in charge of managing the interaction with the environment. This interaction can be considered at two different time scales: on-line interaction and evolution. The on-line interaction refers to the continuous process by which a given individual implemented in the tissue is sensitive to the input stimuli that arrive from the external environment. These stimuli may take the form of any physical magnitude (e.g. light, pressure, or temperature) and, after a conditioning and conversion process, are translated into internal signals that may be used by the individual either to extract some knowledge from the environment or to produce an output as a result of some internal processing. These output signals may be later translated, by means of a set of proper actuators, into physical magnitudes that are reverted as output actions to the

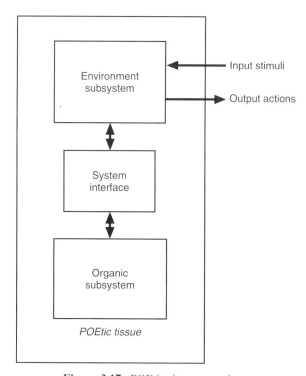

Figure 3.17. POEtic tissue overview.

environment. This on-line interaction constitutes the basic sensor-actuator loop that permits a given individual to adapt its behavior to the specific characteristics of the environment where it is placed. The second kind of interaction with the environment acts at a population level and exceeds the life-time of an individual. In this case the sensor-actuator loop is used to define the basic substrate (the genome) of the individuals that are capable of adapting its behavior to the environment in the most efficient way according to a given fitness measure.

- **Organic Subsystem:** This is in charge of implementing the behavior of an individual, following the principles described by the innate information that has resulted from the evolutionary process. Therefore, it is the goal of this system to permit the development (ontogenesis) of a given functionality from the information stored in a genome, and also to permit the adaptation (epigenesis) of this functionality according to the stimuli received from the environment.

- **System Interface:** This element will allow for an efficient communication between the environment and the organic subsystem of the tissue. It also constitutes the substrate that will provide the basic mechanisms that will permit the scalability of the tissue.

The environment subsystem of the POEtic tissue has been built around a custom 32-bit microprocessor with an efficient and flexible system bus and several custom peripherals. The reason for using a centralized system to carry out evolutionary processes is motivated by the fact that, even if evolution acts on a population of individuals, in the end there must be a global unit that evaluates the fitness of the individuals and determines those from which the next population will be constructed. Therefore, the functionality of the individuals will be implemented in the organic subsystem, but it is the microprocessor that constitutes the core of the environment subsystem that will drive the basic steps of the evolutionary process, as well as the interaction of the individuals with the environment. Additionally, the use of a programmable unit to implement the phylogenetic mechanisms of the tissue will allow for the testing and development of different evolutionary strategies, since this will imply just an update of the software executed by the microprocessor. Finally, this partitioning largely simplifies the management of the acquisition/conversion units that are required to handle the sensor- actuator loop needed to complete the epigenetic and phylogenetic processes to be implemented by the tissue.

Regarding the organic subsystem, one of the main requirements that was considered during its development was the support for the implementation of virtually any type of cell (i.e., a universal cell). To achieve this requirement the organic subsystem is organized around a regular bi-dimensional grid of elementary units that are called molecules. A molecule is a simple piece of configurable digital hardware (a four-input lookup table, a register and a switch box) that may interact with its direct four neighbors. Molecules may be configured to provide a given digital function, and by combining molecules it is possible to construct virtually any type of digital cell. An organism is then able to exhibit a given functionality by combining as many cells as required.

A novel dynamic interconnection mechanism at the cellular level has been designed for the organic subsystem. This strategy permits the cells of an organism to dynamically establish connections during their development process, as well as during their lifetime as obligatory by the interaction with the environment. This dynamic routing strategy, that is absent in classical hardware platforms, is one of the essential elements that will enable the efficient implementation of ontogenetic and epigenetic mechanisms.

The organization of the organic subsystem allows for a layered hardware structure that resembles the principles behind the three main axis of biological organization. At the beginning of the lifetime for a given organism the organic subsystem is just a sea of regular and still unspecialized molecules. During development the molecules organize into cells, and once the cells arise the genome of the organism is interpreted to provide the individual cell configurations required to create a function. After this configuration process is completed, the cells connect themselves autonomously using the dynamic routing principle that has been included in the organic subsystem. At this time the organism shows the innate basic behavior that was determined during the evolutionary process and therefore it is the end of the initial phenotype to genotype translation process. Since cells are composed of configurable elements (the molecules) and they can change their connectivity pattern autonomously at any time, epigenetic processes may take place as required by the specific interaction of the organism with the environment.

POEtic applications are designed around molecules, which are the smallest functional blocks. (What these molecules can do will be discussed shortly.) Groups of molecules are put together to form larger functional blocks called cells. Cells can range from basic logic gates to complete logic circuits such as full-adders. Finally, a number of cells can be combined to make an organism, which is the fully functional application. Figure 3.18 shows a schematic view of the organic subsystem.

Apart from the environment and organic subsystems included in the POEtic architecture, there is a fundamental building block that will permit the tissue to exhibit the inherent features: the system interface. The primary goal of this component is to allow for an efficient interaction between the environment and the organic subsystems. Furthermore, the system interface will cope with the scalability issues of the tissue. Due to technology and cost constraints, when implemented on silicon there is a limit on the amount of resources that constitute the organic subsystem. For this reason, the system interface has to be designed to permit the efficient communication of as many POEtic units as required in order to handle a given application. The system interface has been built around an optimized version of a standard SOC (System On a Chip) bus architecture, the AMBA bus [7].

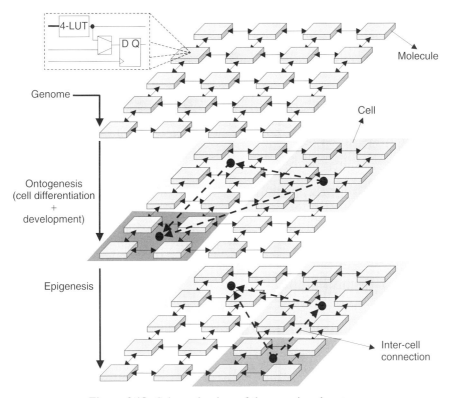

Figure 3.18. Schematic view of the organic subsystem.

Apart from allowing for the construction of an organic subsystem of virtually any size, the system interface allow the construction of an actual distributed sensor/actuator platform, which is common to most living beings.

3.4.1 Organic Subsystem

As explained previously, the organic subsystem is constituted from a regular array of basic building blocks, called molecules, that allow for the implementation of any logic function. This array constitutes the basic substrate of the system. On top of this molecular layer there is an additional "layer" that implements dynamic routing mechanisms between the cells that are constructed combining the functionality of different molecules. Therefore, the physical structure of the organic subsystem can be considered as a two-layer organization (at least in the abstract), as depicted in Figure 3.19.

3.4.2 Description of the Molecules

A molecule is the smallest programmable element of the POEtic tissue. It is mainly composed of a flip-flop (DFF), and a 16-bit look-up table (LUT). Eight modes of operation are supplied to ease the development of applications that need

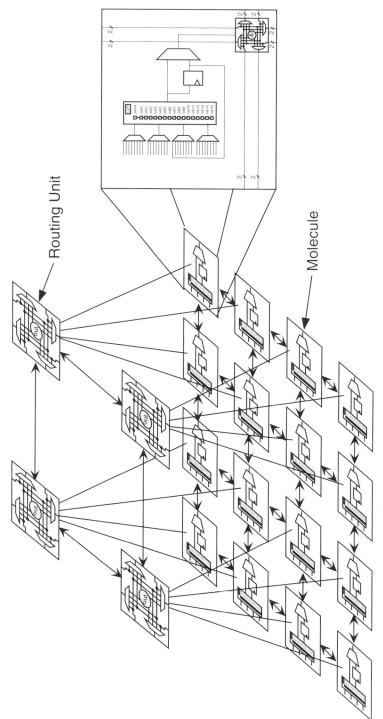

Figure 3.19. Physical structure of the organic subsystem.

cellular systems and/or growth and self-repair. The LUT is composed of a 16-bit shift register that can be split in two, used as a shift register, or as a normal look-up table.

A molecule has eight different operational modes, to speed up some operations, and to use the routing plane. Here we briefly describe the different modes.

- In **4-LUT** mode, the 16-bit LUT supplies an output, depending on its four inputs.
- In **3-LUT** mode, the LUT is split into two 8-bit LUTs, both supplying a result depending on three inputs. The first result can go through the flip-flop, and is the first output. The second one can be used as a second output, and is directly sent to the south neighbor (can serve as a carry in parallel operations).
- In **Comm** mode, the LUT is split into one 8-bit LUT, and one 8-bit shift register. This mode could be used to compare a serial input data with a data stored in the 8-bit shift register.
- In **Shift Memory** mode, the 16 bits are used as a shift register, in order to store data, for example a genome. One input controls the shift, and another one is the input of the shift memory.
- In **Input** mode, the molecule is a cellular input, connected to the intercellular routing plane. One input is used to enable the communication. When inactive, the molecule can accept a new connection, but won't initiate a connection. When active, a routing process will be launched at least until this input connects to its source. A second input selects the routing mode of the entire poetic tissue.
- In **Output** mode, the molecule is a cellular output, connected to the intercellular routing plane. One input is used to enable the communication. When inactive, the molecule can accept a new connection, but it won't initiate a connection. When active, a routing process will be launched at least until this output connects to one target. Another input supplies the value sent to the routing plane.
- In **Trigger** mode, the 16-bit shift register should contain "000...01" for a 16-bit address system. It is used by the routing plane to synchronize the address decoding during the routing process. One input is a circuit enable, that can disable every DFF in the tissue, and another can reset the routing, and so start a new routing.
- In **Configure** mode, the molecule can partially configure its neighborhood. One input is the configuration control signal, and another one is the configuration shifting to the neighbors.

The mode of a molecule is stored in 3 bits of the configuration. They are defined in Table 3.2.

A molecule is defined by 76 configuration bits. Molecules are configured by loading the configuration bits in parallel, from the micro-controller. Partial

TABLE 3.2. Configuration Bits of the Operational Mode

Configuration	Mode
000	4-LUT
001	3-LUT
010	Shift Memory
011	Comm
100	Input
101	Output
110	Trigger
111	Configure

reconfiguration is also possible, a molecule being able to shift configuration bits of its neighborhood. Actually, when shifting a total of 77 bits are used since the value of the flip-flop has to be in the configuration chain in order to retrieve its value.

Long distance inter-molecular communication is possible by the way of switch boxes. Each switch box consists of eight input lines (two from each cardinal direction) and eight corresponding output lines, and are implemented with eight input multiplexers. Two outputs are sent into each of the four neighbors of the molecule, as shown in Figure 3.20.

Each output line can be connected to one of the six input lines from the other cardinal directions (no u-turns allowed) or to one of two possible outputs of the molecules (the output or the inverted output).

3.4.3 Description of the Routing Layer

The second layer of the organism subsystem implements a dynamic routing algorithm to allow the circuit to create paths between different parts of the molecular array. The possibility of having a pseudo-static routing has also been added, to ease the development of applications that only need local connections between cells.

3.4.4 Dynamic Routing

The dynamic routing system is designed to automatically connect the cells' inputs and outputs. Each output of a cell has a unique identifier, at the organism level. For each of its inputs, the cell stores the identifier of the source from which it needs information. A non-connected input (target) or output (source) can initiate the creation of a path by broadcasting its identifier, in case of an output, or the identifier of its source, in case of an input. The path is then created using a parallel implementation of the breadth-first search algorithm. When all paths have been created, the organism can start executing its task. This activity continues until a new routing is launched which occurs, for example, after a cell addition or a cellular self-repair.

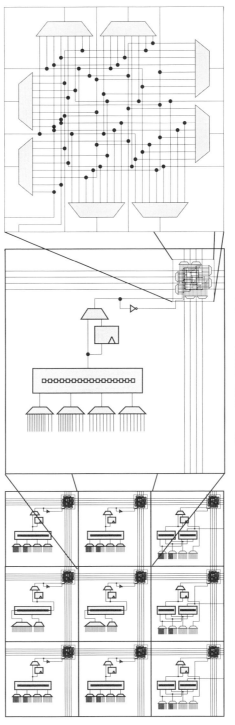

Figure 3.20. POEtic switch box.

This approach has many advantages, compared to a static routing process. First of all, a software implementation of a shortest path algorithm is very time-consuming for a processor, while the parallel implementation requires a very small number of clock cycles to finalize a path. Secondly, when a new cell is created it can start a routing process, without the need of recalculating all paths already created. Thirdly, a cell has the possibility of restarting the routing process of the entire organism, if needed (for instance after a self-repair). Finally, the approach is totally distributed, without any global control over the routing process, so that the algorithm can work without the need of the central microprocessor.

3.5 REMARKS

The choice as to which type of devices to use will obviously depend on the requirements of the end application, both in terms of amount of resources required and the amount of reconfigurability required. However, as technology improves (along the lines of Moore's law) the resource limitation will become less restrictive and so reconfigurability will become the primary factor in choosing an appropriate device.

One product line EHW researchers need to monitor closely are the new FPGA families that provide on-chip processors. For example, the Xilinx Virtex-4 FX platform FPGAs accommodate two PowerPC 405, 32-bit RISC processor cores on a single device. These processors can run the evolutionary algorithm itself—thereby providing an entire EHW digital system on a single COTS chip.

REFERENCES

1. Miller J and Thomson P 2000, "Cartesian genetic programming", *Proceedings 3rd Euro. Conf. on GP (EuroGP2000)*, LCNS 1802, 121–132
2. www.xilinx.com/partinfo/4000.pdf
3. Sekanina L 2003, "Virtual reconfigurable circuits for real-world applications of evolvable hardware", *Evolvable Systems: From Biology to Hardware. Fifth International Conference, ICES 2003*, 186–198
4. Krohling R, Zhou Y and Tyrrell A 2003, "Evolving FPGA-based robot controller using an evolutionary algorithm", *1st International Conference on Artificial Immune Systems*
5. Scott D, Seth S and Samal A 1997, "A synthesis VHDL coding of a genetic algorithm", Technical Report UNL-CSE-97-009
6. Wolfram 1999, "University and complexity in cellular automata", *Physica* 10, 1–35
7. ARM. Amba specification, rev 2.0. Advanced Risc Machines ltd (arm). http://www.arm.com/armtech/amba_spec, 1999
8. Ferguson M, Stoica A, Zebulum R, Keymeulen D and Duong V 2002, "An evolvable hardware platform based on DSP and FPTA" *Proceedings of the Genetic and Evol. Comp. Conf.*, 520–527

9. Shackleford B, Carter D, Snider G, Okushi E, Yasuda M, Koizumi H, Seo K, Iwamoto T, Yasuura H 2000, "An FPGA-based genetic algorithm Machine" *FPGA 2000, Eighth ACM International Symposium on Field-Programmable Gate Arrays*, 218
10. Stoica A, Zebulum R, Keymeulen D, Tawel R, Daud T and Thakoor A 2001, "Reconfigurable VLSI architectures for evolvable hardware: from experimental field programmable transistor arrays to evolution-oriented chips", *IEEE Transactions on VLSI Systems*, 9(1), 227–232
11. Stoica A, Keymeulen D, Thakoor A, Daud T, Klimech G, Jin Y, Tawel R, Duong V 2000, "Evolution of analog circuits on field programmable transistor arrays", *Proceedings of NASA/DoD Workshop on Evolvable Hardware (EH2000)*, 99–108
12. Kajitani I, Hoshino T, Kajihara N, Iwata M, Higuchi T 1999, "An evolvable hardware chip and its application as a multi-function prosthetic hand controller". *Proceedings of 16th National Conference on Artificial Intelligence (AAAI-99)*, 182–187
13. Stoica A, Zebulum R and Keymeulen D 2000, "Mixtrinsic evolution", *Proceedings of the Third International Conference on Evolvable Systems: From Biology to Hardware. (ICES2000)*, 208–217
14. Stoica A, Zebulum R, and Keymeulen D 2001, "Progress and challenges in building evolvable devices", *Third NASA/DoD Workshop on Evolvable Hardware*, 33–35
15. Langeheine J, Becker J, Folling F, Meier K, and Schemmel J 2001, "Initial studies of a new VLSI field programmable transistor array", *Proceedings 4th Int'l. Conf. on Evolvable Systems: From Biology to Hardware*, 62–73
16. Tyrrell A M, Sanchez E, Floreano D, Tempesti G, Mange D, Moreno J M, Rosenberg J and Villa A E P 2003, "POEtic Tissue: An Integrated Architecture for Bio-Inspired Hardware", *Proceedings of 5th International Conference on Evolvable Systems*, Trondheim, 129–140

CHAPTER 4

RECONFIGURABLE ANALOG DEVICES

Aims: *This chapter complements the previous one by describing analog devices. Both commercial and research devices are discussed. A number of experimental results clearly demonstrate the power of analog EHW.*

4.1 BASIC ARCHITECTURES

While digital hardware is becoming more and more powerful, there are a lot of problems requiring analog electronic circuits. An obvious example of this are sensors, that always use some analog front end to measure a physical quantity in an analog world, analog filters or sometimes (massive parallel) signal processing circuits. For the latter example the use of analog circuitry can result in a better ratio of performance and area and/or power consumption. Unlike its digital counterpart, the analog design domain is not blessed with powerful tools that simplifying the design process. This is, at least to some extent, due to the tight relationship between the used technology, the chosen layout and the performance of the resulting circuit all of which makes the simple reuse of standard building blocks without any adaptation virtually impossible. Moreover great care has to be taken in how the specific process parameters can be used to achieve the desired behavior because of the device variations on the actual die. As evolutionary algorithms are assumed to yield good results on complex problems without explicit knowledge of the detailed interdependencies involved, they seem to be a tempting

---

Introduction to Evolvable Hardware: A Practical Guide for Designing Self-Adaptive Systems, by Garrison W. Greenwood and Andrew M. Tyrrell
Copyright © 2007 Institute of Electrical and Electronics Engineers

choice. Accordingly the projects described here try to make a step towards the design automation of analog electronics by means of evolvable hardware.

It is best to begin the discussion with a review of basic analog circuitry. We have already seen that for digital circuits the basic components available to us can be considered as combinational gates (e.g. AND, OR, and NOT) and some simple storage components (e.g. flip-flops). When considering analog circuit design, we have a completely different set of components and issues.

For the purposes of our discussions here, we will consider the basic components available to us for our analog designs as: transistors (P- and N- type), resistors, capacitors, and operational amplifiers. As well as these "high-level" components, with analog technology the actual manufacture of the devices will affect designs: an OR gate is an OR gate and will have the same truth table (functionality) no matter how it has been manufactured (otherwise it would not be an OR gate!). However, we can have an almost infinite number of transistors depending on their manufacture. For example, considering the relationship between the width and length of the channel in a transistor. Also, the same circuit, but with different component values (e.g. resistor values), or with different component types (e.g. a capacitor replacing a resistor) will have significant effects on the characteristics of a design. A few simple examples to illustrate this are shown in Figures 4.1–4.4.

The question we might ask ourselves now is what might an equivalent system look like for analog design *vis-a-vis* a digital FPGA? Given the potentially higher levels of complexity and subtlety within analog designs, there is no one answer to this question. That is, there is no equivalent to the sum-of-products in the analog domain. In an evolvable analog world we might change any or all of the above (and a few other parameters as well). The following sections will give an indication of some of the devices/systems that exist to assist in the evolvable analog hardware process.

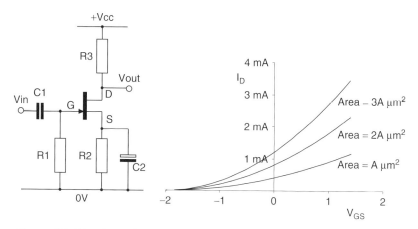

Figure 4.1. Transistor characteristics with varying channel width/length ratios.

BASIC ARCHITECTURES   69

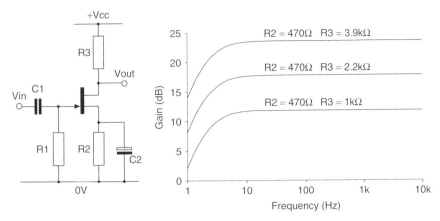

Figure 4.2. Gain vs. frequency characteristics for the amplifier circuit. Curves are shown for three different R3 values.

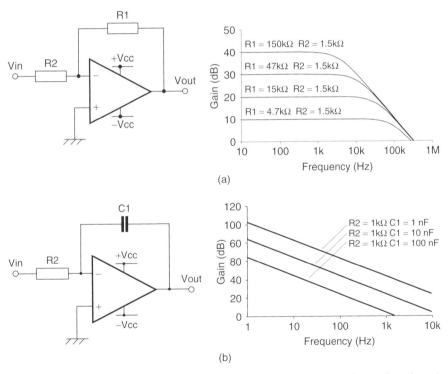

Figure 4.3. Two operational amplifier circuits with resistor (*upper*) and capacitor (*lower*) and associated characteristics.

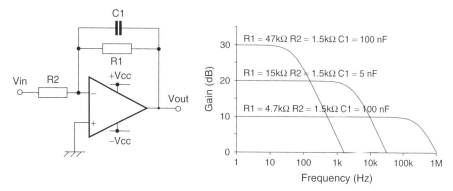

Figure 4.4. An operational amplifier circuit combining the resistor and capacitor circuits shown in Figure 4.3.

4.2 TRANSISTOR ARRAYS

Figure 4.5 gives an overview of three levels of evolvable hardware we might consider. All are real devices and will be considered in more detail in the next few sections. For now however, let's forget they are specific devices and consider them as abstract models for evolvable hardware.

The picture on the left illustrates a device which has a low-level structure where we can vary the actual dimensions of a transistor. We have seen in the previous section where by varying the ratio of the transistor's channel width and length we can obtain transistors with different characteristics. A fine-grained programmable device can be constructed by placing many of these transistor devices into say a $N \times M$ mesh array.

The middle picture moves a level up from this. In this case the transistor dimensions are fixed, which fixes their basic characteristics. A set of these transistors forms a "cell". Now we can program switches between transistors to create different circuits.

Finally, the picture on the right shows a device with blocks of analog circuitry (i.e., operational amplifiers). Here we can not only change the routing between blocks but also select different passive component values within the blocks. Limited configuration changes are also programmable inside the block.

The level of abstraction determines the generic name given to the dives. For example, if the user programs circuitry at the transistor level then that device is referred to as a *field programmable transistor array* (FPTA). At higher levels of abstraction where the user programs say operational amplifier circuitry the device is commonly referred to as a *field programmable analog array* (FPAA).

Having now seen the abstract view of how we might create a platform to allow us to evolve analog systems, let's now consider some specific devices that are available.

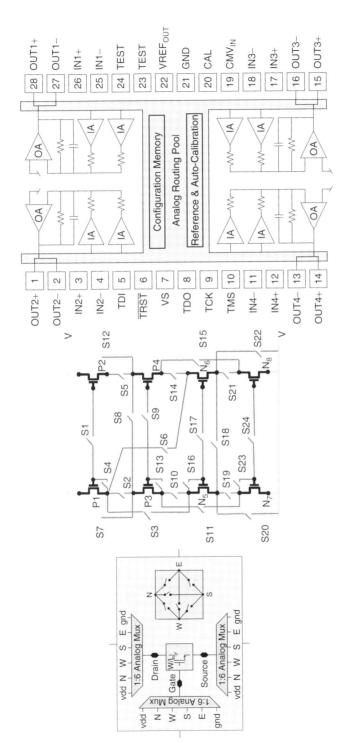

Figure 4.5. Heidelberg one transistor cell: Early JPL transistor array cell: Lattice Semiconductor ispPAC10 modular cell.

4.2.1 The NASA FTPA

The FPTA is an implementation of an evolution-oriented reconfigurable architecture (EORA). The lack of evolution-oriented devices, in particular for analog, has been an important stumbling block for researchers attempting evolution in intrinsic mode (with evaluation directly in hardware). Extrinsic evolution is slow and scales badly when performed accurately (e.g. with SPICE), and less accurate models may lead to solutions that behave differently in hardware than in software simulations.

The FPTA has transistor level re-configurability, supports any arrangement of programming bits without danger of damage to the chip (as is the case with some commercial devices). Three generations of FPTA chips have been built and used in evolutionary experiments. The latest chip, the FPTA-2, consists of an 8×8 array of reconfigurable cells (see Figure 4.6). The chip can receive 96 analog/digital inputs and provide 64 analog/digital outputs. Each cell is programmed through a 16-bit data bus/9-bit address bus control logic, which provides an addressing mechanism to download the bit-string of each cell. Around 5000 bits are used to program the whole chip. The 64 cells can be configured in 1.15 ms (18 μs per cell). The chip has been fabricated using TSMC 0.18/1.8 V (0.18 μm fixture size) technology. The chip die measures 7×5 mm^2, using a total of 256 pins.

Each cell has a transistor array (reconfigurable circuitry shown in Figure 4.7), as well as a set of other programmable resources, including programmable resistors and static capacitors. The reconfigurable circuitry consists of 14 transistors connected through 44 switches and is able to implement different building blocks for analog processing, such as two- and three-stage Operational Amplifiers, logarithmic photo detectors, or Gaussian computational circuits. It includes three capacitors, Cm1, Cm2 and Cc, of 100fF, 100fF and 5pF, respectively.

Each standard cell provides reconfigurable resistances in parallel with the capacitors. Each RC set can be by-passed through a switch. When the reconfigurable circuitry is programmed as an operational amplifier, the "external" resistances and capacitors allow the implementation of inverting amplifiers, integrators, differentiators, etc. The reconfigurable resistances allow programmable flexibility in any configured Operational Amplifier based circuit, and can be programmed to assume the following values: 1.8 k, 6 k, 9 k and 18 kΩ. Four switches control the resistor value (b0 to b3 in Figure 4.8). The two capacitors have values of 100fF.

The pattern of interconnection among cells is similar to that of traditional FPGAs. Each cell interconnects with its north, south, east and west neighbors.

The control logic was designed to allow individual cell configuration. The circuit is shown in Figure 4.9. Each cell requires around 75 bits to be fully programmed. In order to simplify routing, the data bus size was kept to 16-bits. An entire cell therefore needs 5 write cycles (16×5-bits) to be programmed. A decoder of 320 outputs is used to select the cell to be downloaded. Although there are only 64 cells, each one is divided in 5 sub-groups as previously described. A 9-bit address bus is used to configure the decoder.

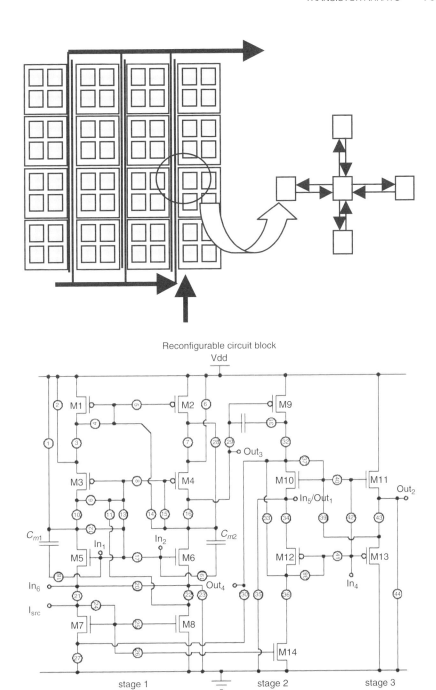

Figure 4.6. The FPTA2 architecture. Each cell contains additional capacitors and programmable resistors (*not shown*).

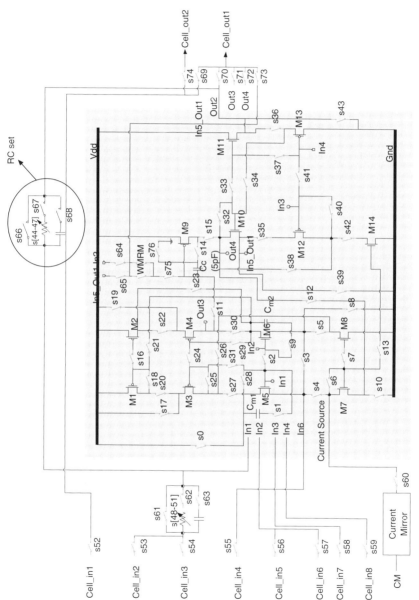

Figure 4.7. Block diagram of the cell.

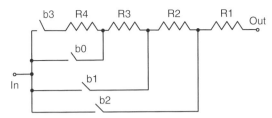

Figure 4.8. Reconfigurable resistance.

Real-world applications will require compact, low-power, autonomous evolvable hardware. A discussion on the evolution of the integration in evolvable systems was presented in [3]. An effort to move from PC-simulated or PC-controlled evolutions to embedded and ultimately to integrated system-on-a-chip evolvable systems is needed. Pioneering efforts in this direction include the work of Higuchi who implemented a genetic algorithm-chip connected to a prosthetic hand [4], and Shackelford who has implemented a digital system [2].

A recent integration effort is the SABLES solution, which provides an autonomous, fast (about 1,000 circuit evaluations per second), on-chip circuit reconfiguration. Its main components are a JPL FPTA chip as transistor-level reconfigurable hardware, and a TI DSP implementing the evolutionary algorithm as the controller for reconfiguration. SABLES achieves approximately 1–2 orders of magnitude reduction in memory and about 4 orders of magnitude improvement in speed compared to systems evolving in simulations, and about an order of magnitude reduction in volume and an order of magnitude improvement in speed (through improved communication) compared to a PC controlled system using the same FPTA chips.

The evolution of a half-wave rectifier circuit is presented to illustrate how the system functions. It also emphasizes the importance of having implicit assumptions explicitly reflected/incorporated in the fitness function. For example, it can be enforced that a logic gate has the same response at different time scales by applying a method derived as a form of mixtrinsic evolution [5] (using a mixed population of individuals tested at two different time scales).

SABLES integrates an FPTA and a DSP implementing the Evolutionary Platform (EP) as shown in Figure 4.10. The system is stand-alone and is connected to the PC only for the purpose of receiving specifications and communicating back the results of evolution for analysis.

The evolutionary algorithm was implemented in a DSP that directly controlled the FPTA, together forming a board-level evolvable system with fast internal communication ensured by a 32-bit bus operating at 7.5 MHz. Details of the EP are presented in [1]. Over four orders of magnitude speed-up of evolution was obtained on the FPTA chip compared to SPICE simulations on a Pentium processor. This performance figure was obtained for a circuit with approximately 100 transistors. The speed-up advantage increases with the size of the circuit and the evaluation time depends on the tests performed on the circuit. Many of

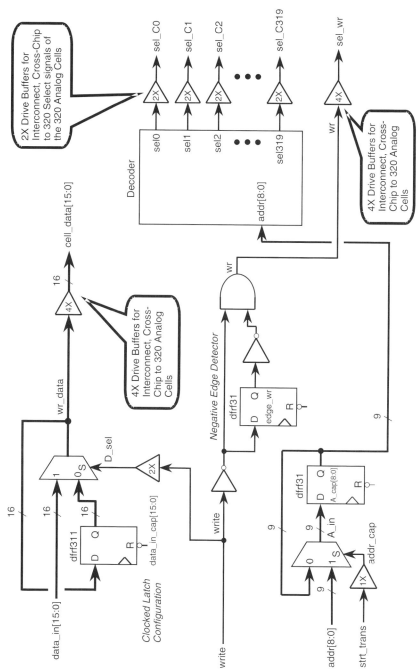

Figure 4.9. FPTA control logic.

Figure 4.10. Block diagram of a simple stand-alone evolvable system.

the evaluation tests performed required less than two milliseconds per individual, which for example on a population of 100 individuals running for 200 generations required only 20 seconds. The bottleneck is now related to the complexity of the circuit and its intrinsic response time.

The following experiment illustrates an evolution on SABLES. The objective of this experiment is to synthesize a half-wave rectifier circuit. The testing of candidate circuits is made for an excitation input of 2 kHz sine wave of amplitude 2 V. A computed rectified waveform of this signal is considered as the target. The fitness function rewards those individuals exhibiting behavior closer to the target (using a simple sum of differences between the response of a circuit and target) and penalizes those further from it. After evaluation, all individuals are sorted according to fitness. Elitism is used with the highest fit 9% of the population preserved. The remaining individuals initially undergo crossover (with a 70% probability), and then mutation (with a 4% probability). The entire population is then re-evaluated. In this experiment only two cells of the FPTA were allocated.

The left side of Figure 4.11 illustrates the programming time of the new circuit and the stimulation with two waveform periods with two different looking responses for the two circuits being evaluated. The right side of Figure 4.11 depicts the waveforms for stimulus and response and the time allocated for stimulation and the time allocated for the GA in an evolutionary cycle.

Figure 4.12 displays snapshots of evolution in progress, illustrating the response of the best individual in the population over a set of generations. The first caption shows the best individual of the initial population, while the subsequent ones show the best after 5, 50 and 82 generations respectively. The solution, with a fitness below 4,500, is shown on the right of the figure. Figure 4.13 shows the convergence over a number of runs.

With SABLES enabling rapid evolvable hardware experiments, the focus has shifted from the hardware platform to algorithms. More specifically the focus became overcoming problems related to the formulation of requirements in a way that facilitate evolutions, and the translation of target specifications into the language of evolution, including representations, fitness function and parameters of the algorithm. One reason for the difficulty of evolving autonomous systems is the need to provide complete specifications. It should be emphasized that completeness may not often be obvious. In computer-assisted design the human can come back and provide extra information that may have been omitted in the beginning. That is not possible in an autonomous system, and evolution usually

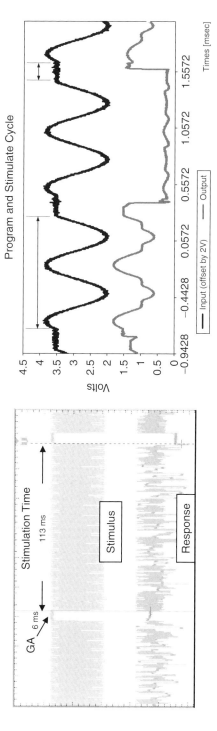

Figure 4.11. Stimulus-response waveforms for 2 individuals in the population (*left*) and during the evaluation of a population in one generation (*right*). A full GA cycle includes stimulus/response (113 ms) and the generation of the next generation (6 ms). The response was sampled at the maximum sampling rate of the on-board A/D (100 KS/sec) [1].

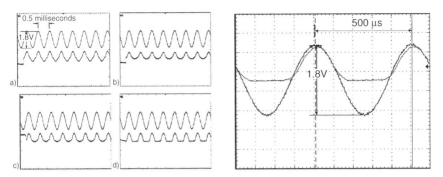

Figure 4.12. Evolution of a half-wave rectifier showing the response of the best individual of generation a) 1, b) 5, c) 50 and finally the solution at generation d) 82. The final solution, which had a fitness value less than 4500, is illustrated on the right [1].

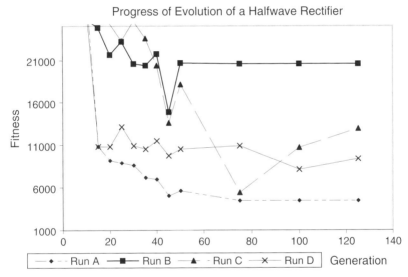

Figure 4.13. The fitness function against generations. The first few generations had fitness values near 100,000, which are not shown on this scale [1].

finds the easiest way to satisfy expressed requirements. It would cheat if it can find it an evolutionary advantage!

Transient Solutions

The half-wave rectifier experiment provides examples of two situations in which evaluations of the candidate solutions on the same hardware-in-the-loop may lead to highly-ranked individuals receiving high fitness values and yet, when re-evaluated individually, prove to have been only spurious solutions. Figure 4.14

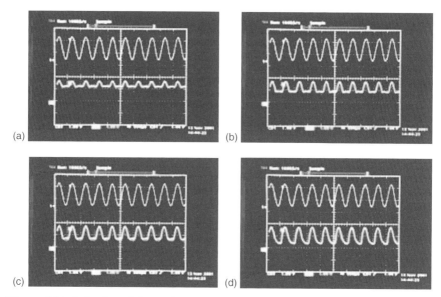

Figure 4.14. Behavior of the evolved half-wave rectifier. The degradation shown from (a) to (d) occurred over the span of approximately 1 second [1].

illustrates both a transient behavior and FPTA-state dependence. The transient behavior describes a configuration that is not stable as a function of time, whereas FPTA state dependence describes a configuration whose behavior depends on the previous configuration(s). Both of these behaviors are shown in Figure 4.14; most obviously, the function, which starts out looking similar to a half-wave rectifier ends up looking quite different. The transient behavior in this case occurred on a time-scale of around 1 second. Despite the transient behavior, the individual was selected as a solution because it was evaluated during a time-scale of around 2 milliseconds much shorter than the transient duration. In practice the transient behavior can be resolved by re-evaluating the individuals for a longer time period.

The FPTA state dependence behavior occurred when the individual solutions programmed on the FPTA suffered somewhat from an apparent instability, which arose when the evaluation of a given individual depends on the previous state of the FPTA. The individual shown in Figure 4.14 was selected as a solution because, during the evaluation, its response matched quite well the expected function. However part a) of Figure 4.14 shows that the circuit does not behave sufficiently like the target rectifier, so the behavior exhibited in the evaluation must have been influenced by the previously downloaded configuration(s).

Parasitic as well as static capacitors in the chip explain this behavior. Capacitors can be charged during one configuration period and not be completely discharged before the next configuration is tested, which leads to an undefined charge on capacitors and subsequently alters the behavior of the circuit. Nevertheless, for this particular experiment it has been observed that evolution weeds

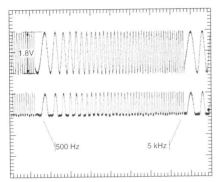 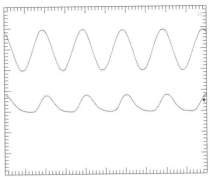

Figure 4.15. Response of the half-wave rectifier for a frequency sweep from 500 Hz to 5 kHz (left). Deteriorated response at 50 kHz [1].

these individuals out and stable solutions are almost always found within the first 50 generations, or about 20 seconds.

The operational range of the evolved circuit in the frequency domain is another potential pitfall, since in principle the circuit behavior should be evaluated for the overall frequency domain in which it is expected to work. Figure 4.15 depicts the half-wave rectifier response for different frequencies. From the graph on the left of the figure, it can be observed that the circuit works correctly for the decade going from 500 Hz to 5 kHz, the frequency region in which the circuit was actually evolved. Nevertheless, the response deteriorates for higher frequencies, as illustrated by the graph at the right of Figure 4.15 for 50 kHz.

These results are typical of a series of successful runs. Approximately 1 out of 10 runs ended with the algorithm getting stuck and not finding a solution at all for that small/fixed mutation rate.

Time Constants

Some of the assumptions often implicit to human designers may be missing from the explicitly formulated requirements and thus from the fitness function of an evolutionary design. An example illustrated here is the implicit assumption that a logic gate should have the same behavior over a "frequency range" i.e. function with slow/DC signals as well as to faster changing input signals. The example illustrated here evolved a NAND gate where the input stimulus (using a SPICE transient analysis) changes are in the microsecond range. For this timescale evolution quickly created a properly functioning logic gate. However, the NAND gate exhibited incorrect behavior when simulated in the timescale of seconds. Referring to Figure 4.16, the first column shows the response of the evolved circuit at the scale used in the fitness function (microseconds), while the second column shows the response of the same circuit when evaluated at a different timescale (seconds). Conversely, when the circuit is evaluated at a large timescale evolution often led to slow gates. The method applied to correct this situation was a derivative of the mixtrinsic evolution method introduced in [5].

In mixtrinsic evolution two type of models, which could be either software or hardware, are used to compute a solution's fitness. In some cases these models are subject to different types of evaluation, which is what was done here. There are two ways of computing the fitness: (1) with a combined fitness function for the two models, or (2) by assigning candidate solutions to a different model during successive generations and letting evolution remove solutions that do not behave well with both models.

In this particular experiment a two-transient analysis for each candidate circuit was evaluated first on a small timescale and second on the larger timescale. The combined fitness measure was the worse performance between the two evaluations thereby forcing the genetic algorithm find a circuit that exhibited correct behavior at both timescales. The two right columns of Figure 4.16 show the response of a circuit evolved using this method. Notice that a correct response was achieved at both timescales.

Comment: The FPTA components and the stand-alone board-level evolvable system (SABLES) provide a fast, and flexible compact stand-alone evolvable system for both analog and digital circuits. While such systems have advantages there are also lessons to learn when using such systems and techniques. Intrinsic evolution using the same hardware-in-the-loop resources for consecutive evaluation of individuals may lead to transient solutions. In most cases these can be eliminated simply by allowing extra time for evolution. An additional possible trap may come from incomplete requirement specification such as those related to the timescale of operation of logic gates, in which case either slow or gates only operating on fast changing inputs may be obtained. These can be solved through mixtrinsic evolution, i.e., with a combined fitness function to reflect desired behavior at two timescales.

4.2.2 The Heidelberg FPTA

The motivation for the design of the Heidelberg FPTA[1] is threefold: First, it is designed as a search tool to find new analog transistor level circuits. Accordingly, the cells of the FPTA are used as a model for programmable CMOS transistors such that evolved circuits can be understood in terms of simulation and human design experience. The use of hardware-in-the-loop may accelerate the evaluation of candidate solutions while avoiding the simplifications inherent to simulations. Second, the FPTA is a step towards field evolvable hardware. The device may be used to perform analog tasks that cannot be *a priori* specified or need the analog circuit to adapt to changing environments. Third, the FPTA can be used as a research tool to learn how to use artificial evolution for the invention of systems with higher complexity from an algorithmic point of view.

The FPTA consists of 16×16 programmable transistor cells. As CMOS transistors come in two flavors, namely N- and P- MOS, half of the transistor cells are designed as programmable NMOS transistors and half as programmable PMOS transistors. P- and N- MOS transistor cells are arranged in a checkerboard pattern as depicted in Figure 4.17.

---

[1] Although Heidelberg devices and JPL devices are both called FPTAs, the two devices were developed independently.

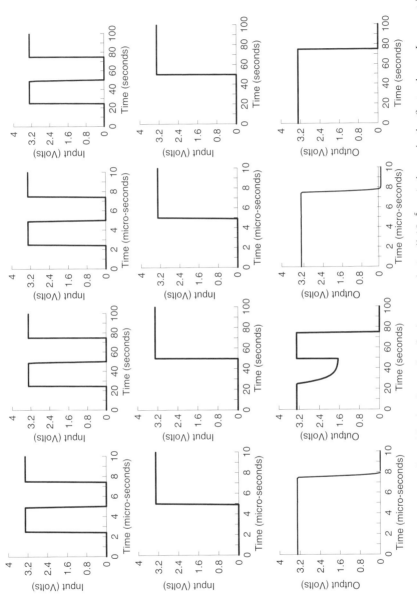

Figure 4.16. Evolved NAND gate evaluated in the timescale of microseconds (until 10^{-5} sec) shown in the first column. Incorrect behavior of the gate when it is simulated in the timescale of seconds (until 100 seconds) is shown in the second column. Evolved NAND gate using two different timescales (micro-seconds in the left and seconds in the right) show a correct behavior in columns three and four [5].

83

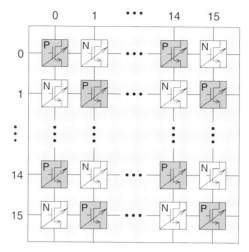

Figure 4.17. Schematic diagram of the FPTA.

Each cell contains the programmable transistor itself, three decoders that allow the three transistor terminals to be connected to one of the four cell boundaries, V_{dd} or Gnd, and six routing switches. A block diagram of the transistor cell is shown in Figure 4.18. Width W and Length L of the programmable transistor can be chosen to be 1, 2, . ., 15 μm and 0.6, 1, 2, 4, or 8 μm respectively. The three terminals drain, gate and source of the programmable transistor can be connected to either of the four cell edges named after the four cardinal points, as well as to V_{dd} or Gnd. The only means of routing signals through the chip is given by the six routing switches that connect the four cell borders with each other. Thus, in some cases it is not possible to use a transistor cell for routing and as a transistor. More details on the FPTA can be found in [6].

At each corner some of the configuration information is stored in a block of static RAM containing 6 bits each. Of the 22 bits used, 6 bits directly control the routing switches that route signals through the cell. Each terminal of the programmable transistor, whose channel geometry is set by 7 bits, can be connected to either power (V_{dd}), ground (Gnd) or any of the four edges of the cell, named after the four cardinal points. The remaining two codes of the multiplexers for drain and source are used to leave the terminals floating. For the gate the same code ties the gate terminal to power or ground for P- and NMOS transistors respectively, thus disabling the transistor.

To analyze the behavior of successfully evolved circuits, the voltage at the east, south, drain and source nodes can be read through a unity gain buffer. Consequently, all the nodes between adjacent cells can be read out and all currents through the active transistors can be estimated by the voltage drop across the transmission gates connecting them to the cell borders.

The evolution system, illustrated in Figure 4.19, can be divided into three main parts: the actual FPTA chip serving as the silicon substrate to host the

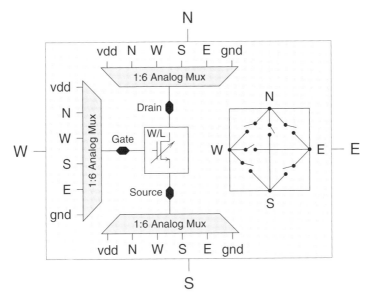

Figure 4.18. Simplified schematic of one transistor cell.

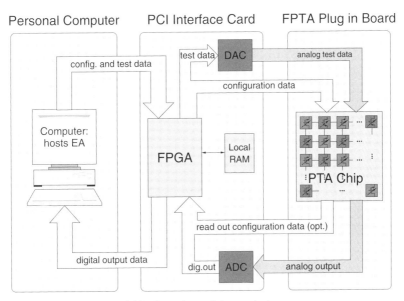

Figure 4.19. Overview of the evolution system.

candidate circuits, the software that contains the search algorithm running on a standard PC and a PCI interface card that connects the PC to the FPTA chip. The software uploads the configuration bit strings to be tested to the FPTA chip via the PCI card. In order to generate an analog test pattern at the inputs of the

FPTA chip, the input data is written to the FPGA on the PCI interface card. There it is converted into an analog signal by a 16-bit DAC. After applying the analog signal to the FPTA, the output of the FPTA is sampled and converted into a digital signal via a 12-bit ADC. The digital output is then fed back to the search algorithm, which in turn generates the new individuals for the next generation. Below we describe the evolution of quasi-DC behavior to illustrate how the evolution system performs [6].

Analog circuits are usually simulated and tested using different test modes. For instance, while DC sweeps test the static behavior of the circuit, transient and AC analyses are used to characterize time and frequency response of the device under test. It is therefore important to randomized the test patterns to prevent the candidate solutions from using any temporal correlation among these test inputs. This sort of DC test was used to evolve the quasi DC behaviors of logic gates and Gaussian voltage transfer characteristics (V-V curves).

Throughout all of the following experiments a simple GA with truncation selection was used. The circuit representation in the genotype preserves the transistor cell structure of the FPTA. In other words, while the mutation operator is free to change any characteristic of any transistor cell, crossover is restricted to exchanging two-dimensional blocks of transistor cell data. The quadratic deviation of the measured circuit response from the desired target behavior is taken as a fitness criterion.

Any deviation constitutes an error so the objective is to minimize it during the evolution run. In the results that follow the errors expressed as a root mean square error (in mV) per measured input data point. That is,

$$\text{RMS Error} = 1000 \cdot \sqrt{\frac{\sum_{i=1}^{512}(V_{tar}(i) - V_{out}(i))^2}{512}},$$

with the target voltage being a function of two input voltages in the more general case of the logic gates experiments

$$V_{tar} = V_{tar}(V_{in1}, V_{in2}).$$

In both experiments the generation size was set to 50 and 512 test points were used. The logic gate experiments ran for 5000 and the Gaussian output characteristic experiments for 10000 generations, these runs took about 30 and 60 minutes respectively [6].

Logic Gates

In a series of artificial evolution experiments the quasi DC behavior of the six symmetric logic gates NOR, NAND, AND, OR, XOR and XNOR was evolved. A total of 100 runs were performed for each of the gates. The successful solutions were required to respond with the correct output voltage (5 V for a logic one and 0 V for a logic zero) depending on the two input voltages. The input was considered low if it was below 1.7 V and high for voltages higher than 3.3 V. During all runs only 5 × 5 transistor cells were accessible for the GA.

As can be seen from the histograms in Figure 4.20, the difficulty to find good solutions for the different gate types corresponds to the different levels of complexity exhibited by the appropriate textbook solutions: While good solutions to the NOR and NAND problem are found quite frequently, the success rate for the OR and AND gates is considerably lower. In 100 runs, no perfect solution could be obtained for the XOR gate or for the XNOR gate.

Figure 4.21 compares the measured behavior of the best NAND, AND and XNOR gates evolved to the simulated characteristics of their textbook counterparts. The simulation results were obtained using the same CMOS process the FPTA was fabricated with. The evolved circuits were tested outside of the optimization loop to analyze the stability of the evolved solutions. The tests were repeated on a second FPTA chip. The vast majority of the evolved circuits and in particular the successful ones from Figure 4.21 are based on achieved

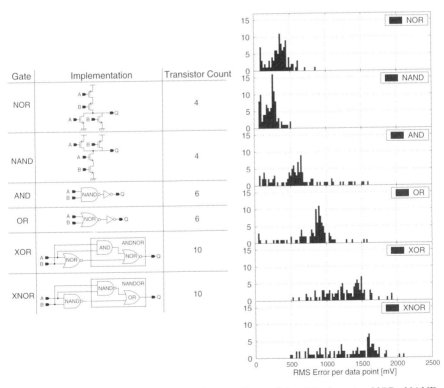

Figure 4.20. *Left*: Typical CMOS implementations of the 6 logic gates NOR, NAND, AND, OR, XOR and XNOR. *Right*: RMS deviation from the ideal response for 100 evolutions of the DC behavior of these 6 logic gates [6].

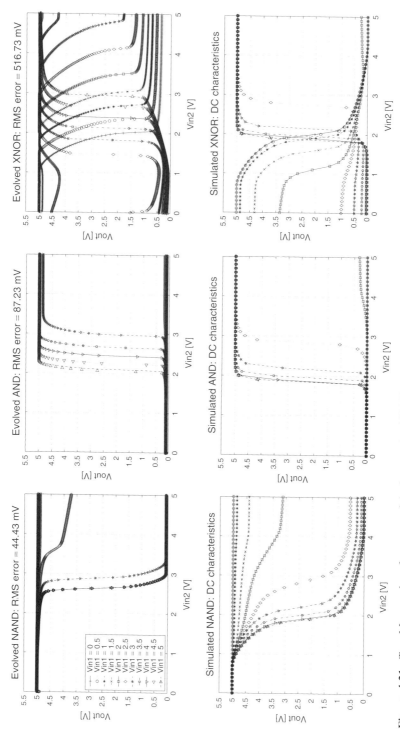

Figure 4.21. *Top*: Measured performance of the best evolved NAND, AND and XNOR gates. *Bottom*: Simulation results for the NAND, AND and XNOR gates depicted on the left side of Figure 4.20. The legend shown in the plot in the upper left corner is used for all 6 plots [6].

similar RMS errors on both chips, which indicates a minimum of robustness against the device fluctuation inherent to the fabrication process of the chip.

By means of similar experiments focusing on finding symmetrical logic gates the influence of different test pattern application schemes has been investigated. The results indicate that it is a) necessary and b) sufficient to apply the input value pairs in random order to prevent the evolution process from abusing time dependencies inherent to the test patterns. However, this procedure cannot rule out long-term instabilities occurring on time scales larger than the ones used during the evaluation of the circuits.

Gaussian Output Characteristics

In a second series of experiments attempt was made to evolve a circuit that has a Gaussian DC characteristic. Different experiments are carried out, in which the fraction of the chip that can be used by the evolutionary algorithm is varied. For each edge length of the quadratic array available to the GA 10 runs are carried out. The edge lengths range from 4 to 11. Figure 4.22 shows the output characteristics of the best evolved solutions for edge lengths between 4 and 9 cells. On the one hand, many of the evolved circuits match the required output voltages quite closely, but on the other hand, the resulting curves do not exactly follow a Gaussian shape.

Being restricted to the same amount of computing time, runs having access to more resources do not necessarily produce better solutions. However, the RMS error for the best solution found for edge length 4 seems to be a bit worse than the best circuits of the other edge lengths.

The Heidelberg FPTA project is one of the few approaches to intrinsic transistor level hardware evolution. Since the chip is based on programmable transistor cells it can be used to find new circuit topologies. However, the configurability of the FPTA entails additional parasitic resistances and capacitances that limit the maximum bandwidth of the transistor array and slightly deteriorate the transistor cell characteristics.

The two experiments presented show the capability of the system to find solutions to well known problems as well as to less common tasks. The fact that most of the evolved circuits performed equally well on two different die outside of the optimization loop indicates that they do not rely on details of the particular transistor cells they have been evolved on.

4.3 ANALOG ARRAYS

The Lattice Semiconductor ispPAC devices are typical of what is currently available at a higher level of complexity than that of the transistor arrays. The isp family of COTS devices provide three levels of programmability: the functionality of each cell, the performance characteristics for each cell, and the interconnect at the device architectural level. Programming, erasing, and reprogramming is achieved quickly and easily through standard serial interfaces while the devices remain soldered to the printed circuit board.

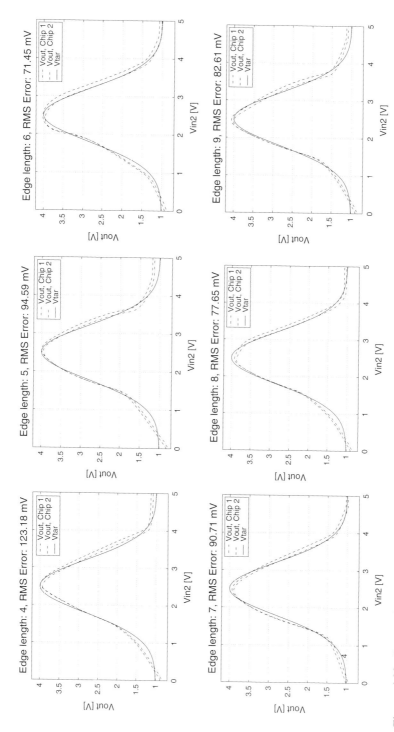

Figure 4.22. Output characteristic of the best Gaussian circuits for different available array sizes. Chip 1 refers to the chip the circuits were evolved on. The curves denoted with chip 2 illustrate the response of the same circuits tested on a different die [6].

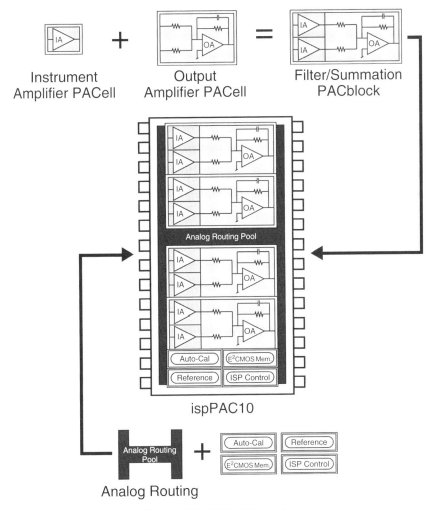

Figure 4.23. PACell hierarchy.

The device hierarchy is depicted in Figure 4.23. The basic active functional element of the ispPAC devices is the PACell which, depending on the specific device architecture, may be an instrumentation amplifier, a summing amplifier or some other elemental active stage. Analog function modules, called PACblocks, are constructed from multiple PACells to replace traditional analog components such as amplifiers and active filters, eliminating the need for most external resistors and capacitors. Requiring no external components, ispPAC devices flexibly implement basic analog functions such as precision filtering, summing/differencing, gain/attenuation and conversion.

One key element to achieving the flexible integration on the ispPAC devices is the Analog Routing Pool (ARP). The ARP provides a programmable analog wiring network within the ispPAC devices between device pins and the inputs and outputs of PACells and PACblocks. The ARP allows these building blocks to be individually connected to the appropriate input or output pins as well as allowing stages to be cascaded without external connections for more complex circuit arrangements.

The specific number of PACblocks and their functions varies from ispPAC device to ispPAC device to provide optimized application coverage and wider designer options. For example, the ispPAC10 device provides four identical PACblocks optimized for signal conditioning and filtering, while the ispPAC20 provides two similar PACblocks plus DAC and Comparator PACells offering conversion and monitoring capabilities. Each PACblock contains a summing amplifier and two differential input instrument amplifiers, as well as an array of feedback capacitors. These capacitors provide the user with programmable pole frequencies to tune a circuit's frequency response Variable gain input instrument amplifiers make it possible to program in steps between ± 0.1 and ± 1.0 and in integer steps between ± 1.0 and ± 10. More complex signal processing functions are performed by configuring and interconnecting additional PACblocks with each other to achieve a variety of circuit functions. This ispPAC10 example only hints at the programmable capabilities of ispPAC devices, an example of using these devices will be given in Chapter 5.

Just to give a little more detail, the PACblock model from PAC-Designer is shown in Figure 4.24. The output amplifier is configured as an inverting mode op amp and illustrates the summing configuration. The input instrument amplifiers are shown to make it clear that unlike a typical inverting op amp, the PACblock input impedance is extremely high. The input amplifier (IA) transconductance gain is shown as the value "k" above or below each amplifier. The gain of IA1 and IA2 is independently programmable. The feedback transconductor IAF (designated here as R_F) can be disabled by the user. Hence, a user configurable switch is shown in series with R_F.

Up to 128 different feedback capacitor values can be chosen. Notice the capacitor is permanently installed so the circuit naturally implements a first-order

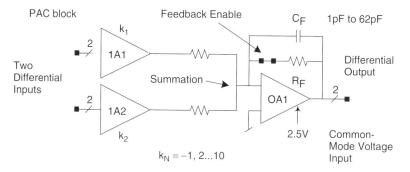

Figure 4.24. PACBlock model.

lowpass filter when the feedback enable is closed or an integrator when it is open. If a simple inverting amplifier is needed C_F is set to 1 pF to put the pole at the highest possible frequency.

While nothing we have said, so far, relates directly to evolvable hardware, the important point here is the ispPAC family of devices are programmable and give us an opening in the analog domain to designing evolvable systems with COTS devices, in a similar way to what we have seen with FPGAs in the digital domain. However, there are limitations that designers need to consider. That issue discussed in Chapter 5.

As with digital devices, we have considered a number of practical aspects of analog devices that might be used in the design and construction of reconfigurable systems. Once again more will be seen of these devices in Chapter 5, when further applications will be illustrated and other appealing features of these systems will be explored. For now however, that is all we want to say about analog devices.

4.4 REMARKS

We have now demonstrated that analog systems are ideal venues for reconfigurable devices. The transistor arrays are the most flexible, but unfortunately no COTS versions are now available—nor likely to be available in the near future. This has nothing to do with their utility, but everything to do with marketplace forces. FPTAs are presently restricted to a small number of niche application areas in which they are superb choices for a reconfigurable device. Nevertheless, the small number of applications also means their demand in the marketplace isn't sufficient to warrant a commercial product line. We conjecture FPTA-type devices will be restricted to military and space applications. This means future FPTA development most likely will be done internally by government agencies or by companies in the defense or aerospace industry with no intention of introducing a COTS version.

The analog array market should remain strong primarily because of an increasing demand for mixed-signal applications. However, the current COTS devices are, in our opinion, overly restrictive in what on-chip resources are reconfigurable. Much of the basic structure is fixed by the manufacturer and therefore not programmable. Do not misunderstand what is being said here. The analog COTS devices available now are quite powerful and perfectly suitable for real-world EHW applications. We would just like to see more reprogrammable on-chip resources made available, which will open up even more application areas.

REFERENCES

1. Ferguson M, Stoica A, Zebulum R, Keymeulen D and Duong V 2002, "An evolvable hardware platform based on DSP and FPTA" *Proceedings of the Genetic and Evol. Comp. Conf.*, 520–527

2. Shackleford B, Carter D, Snider G, Okushi E, Yasuda M, Koizumi H, Seo K, Iwamoto T, Yasuura H 2000, "An FPGA-based genetic algorithm Machine" *FPGA 2000, Eighth ACM International Symposium on Field-Programmable Gate Arrays*, 218
3. Stoica A, Keymeulen D, Thakoor A, Daud T, Klimech G, Jin Y, Tawel R, Duong V 2000, "Evolution of analog circuits on field programmable transistor arrays", *Proceedings of NASA/DoD Workshop on Evolvable Hardware (EH2000)*, 99–108
4. Kajitani I, Hoshino T, Kajihara N, Iwata M, Higuchi T 1999, "An evolvable hardware chip and its application as a multi-function prosthetic hand controller". *Proceedings of 16th National Conference on Artificial Intelligence (AAAI-99)*, 182–187
5. Stoica A, Zebulum R and Keymeulen D 2000, "Mixtrinsic evolution", *Proceedings of the Third International Conference on Evolvable Systems: From Biology to Hardware. (ICES2000)*, 208–217
6. Langeheine J, Becker J, Folling F, Meier K, and Schemmel J 2001, "Initial studies of a new VLSI field programmable transistor array", *Proceedings 4th Int'l. Conf. on Evolvable Systems: From Biology to Hardware*, 62–73

CHAPTER 5

PUTTING EVOLVABLE HARDWARE TO USE

Aims: *This chapter shows how EHW is used for circuit synthesis and hardware adaption. A number of examples, both analog and digital, show the basic techniques. Some background in fault tolerant systems is provided before describing how EHW methods are used for fault recovery; Section 5.2 should be sufficient for most readers.*

5.1 SYNTHESIS VS. ADAPTION

Chapter 1 gave a basic overview of evolvable hardware (EHW). Two main application areas were identified: *circuit synthesis* and *adaptive hardware* In this chapter we describe how EHW does this.

A good place to start is to revisit the description of the components in an EHW system. Essentially we can think of EHW as

EHW = reconfigurable hardware + reconfiguration method.

Notice we used the term "reconfigurable hardware" instead of "reconfigurable devices". That's because reconfiguration is not restricted to just devices like FPGAs. For example, Linden [59] used EHW techniques to design antennas for wireless communications systems. The reconfiguration method must intelligently search for a good configuration. Almost exclusively this method is some type of evolutionary algorithm. At present this EA runs external to the reconfigurable

Introduction to Evolvable Hardware: *A Practical Guide for Designing Self-Adaptive Systems*,
by Garrison W. Greenwood and Andrew M. Tyrrell
Copyright © 2007 Institute of Electrical and Electronics Engineers

hardware, but the increasing power of systems-on-a-chip suggests in the near future the EA could run on the same platform.

Circuit synthesis is, in principle, relatively straightforward: simply evolve a circuit that satisfies some given specification. In practice this evolution is difficult because the search space of all possible circuit configuration is enormous—especially if evolving the circuit topology is also being done. Adaptive hardware applications are a bit more complicated because in most cases reconfiguration is done on faulty hardware or in a hostile and largely undefined operational environment. Under such conditions it is not possible to know *a priori* what type of performance evolution can achieve. Nevertheless, evolving a new circuit configuration may be the only viable means of restoring some operation.

The evolution can be done intrinsically (in hardware) or extrinsically (in software). Much of the circuit synthesis work done today relies on simulators. Unfortunately simulators do not always scale well, which says large designs may have to be done intrinsically to limit the design time. Adapting hardware has to be done intrinsically. (This aspect is discussed in depth later in the chapter.)

The principles underlying circuit evolution are exactly the same whether a circuit is being synthesized or being adapted. However, adapting a circuit has some special considerations that require the designer to have a foundation in fault tolerant concepts and real-time systems. (This foundation is provided in the next section). A number of examples are provided in this chapter, which should be sufficient to show the reader how hardware is evolved.

5.2 DESIGNING SELF-ADAPTIVE SYSTEMS

Self-adaptive systems automatically modify their behavior to compensate for failures or environmental changes. These modifications will only be effective if they take place within specified timeframes. Hence, self-adaptive systems must have both fault tolerant and real-time properties. These types of systems are described in this section. Further information can be found in [54].

5.2.1 Fault Tolerant Systems

All systems eventually fail. In some cases failures cause no perceptible affect on the system's behavior. For example, an input clamping diode that suddenly opens will have no effect so long as the input voltage stays within prescribed bounds. In other cases failures have very perceptible effects that can range from mild annoyance to total system destruction. The best way to improve system availability is to make the system *fault tolerant*—that is, a system that can continue operating, albeit with degraded performance, in the presence of failures. One fault tolerant method masks failures by producing multiple independently generated results and then taking a majority vote. But hiding a failure does nothing to correct the failure. This can have severe consequences: left untreated, some failures can propagate thereby inducing even more failures.

The best solution is to fix the failures as quickly as possible. Designers must therefore anticipate how failures can occur, incorporate features into the design to mitigate any failure effects and, if possible, repair the system. Although fault masking qualifies as a fault tolerant method, we restrict the designation "fault tolerant" to those systems that can autonomously—that is, without human intervention—detect and recover from hardware faults. Fault-tolerance under this interpretation involves two distinct processes: *fault detection and isolation* (FDI) and *fault recovery*. EHW has no real role in FDI but it is perfectly suited for conducting fault recovery operations. However, since FDI techniques are used to evaluate recovery methods, for completeness this section describes both FDI and fault recovery principles.

5.2.1.1 Fault Detection and Isolation FDI methods try to determine if a system's behavior has changed, and if so, identify the specific component that failed. Fault detection, in principle, is easy: just observe the system response and see if it performs as expected. Normally there are some performance thresholds established and exceeding the threshold indicates a system failure. However, some care is needed when setting the thresholds: if the threshold is too wide, a failure may go undetected; if the threshold is too tight, false positives can occur. Actually, the real difficulty is fault isolation. What makes fault isolation hard is more than one potential failure can often produce the same symptoms. For instance, suppose samples from a temperature sensor are run through an analog-to-digital converter that is physically connected to a computer's input port. Software running on the computer reads the value and then stores it in memory. In this system any erroneous temperature readings could be due to an analog-to-digital converter failure, an input port circuitry failure, or even a software failure. It will be impossible to restore normal system operation until the source of the failure is known.

As with any field, understanding depends on a correct use of terminology. We therefore begin with some key definitions.

Definition: (*mishap*)

> Any unplanned event or series of events resulting in injury, death, or some other catastrophic condition such as explosion, fire or system destruction.

Definition: (*hazard*)

> Any real or potential condition that can cause injury, illness or death to personnel; system damage or destruction: or environmental damage.

Definition: (*failure*)

> The inability to accomplish an assigned task.

Definition: (*fault*)

> A defect that can lead to a failure.

98 PUTTING EVOLVABLE HARDWARE TO USE

There is a causal relationship between these terms. Specifically,

fault → failure → hazard → mishap.

Although the example shown in Figure 5.1 is not an EHW problem, it does show how these terms are related to each other.

It is often not possible to make a system mishap-free. In other words, there is always some level of mishap risk and there may be little a designer can do about it. (There is no way to guarantee a car will never be in an accident. However, the

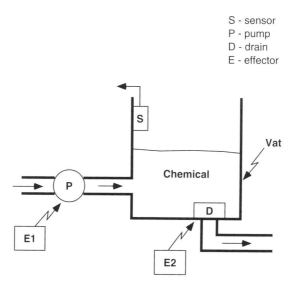

S - sensor
P - pump
D - drain
E - effector

Effector E1 failure: pump won't turn off
Effector E2 failure: drain won't open

Sensor failure: can't tell when vat is full

fault: transistor shorts in E2
failure: drain won't open
hazard: can't empty vat
mishap: liquid exposed to air which releases toxic gas

Figure 5.1. The vat is used to mix several toxic chemicals. The entire system is under computer control. Effectors are circuitry that converts computer commands to physical signals needed to control the plant machinery. Specifically, effector E1 turns the pump on and off (to fill the vat) and effector E2 opens and closes the drain (to empty the vat). Sensor S indicates when the vat is full. Potential failures for E1, E2 and S are indicated. An example of a fault that leads to a mishap is shown in the figure.

risk is negligible with a good driver.) Some examples of hazards include complete loss of flight control in a spacecraft or the release of a hazardous material such as a toxic gas or liquid. Failures do not always cause hazards. For example, diagnostic circuitry failures may no effect on a system's performance when it is in normal operating mode. In most cases, however, failures result in undesirable system behavior that ranges from annoying to unacceptable; the change in behavior is serious enough that something, such as fault recovery measures, must be done to correct the problem.

Definition: (*failure mode*)

> A type of component or system defect that adversely affects behavior or performance.

The objective is now to find out how the system can fail and the resultant effects. This evaluation can be done in several ways and the interested reader can find more comprehensive information elsewhere (e.g., Dunn [49] contains an excellent introduction). Systems are typically evaluated with one of the following methods:

- *failure modes and effects analysis* (FMEA) A bottom-up method that looks at each component's failure modes and how those failures affect system performance.
- *fault tree analysis* (FTA)
  A top-down method that assumes a specific failure has happened and then analyzes each subsystem to look for the cause.
- *failure modes and effects testing* (FMET)
  Failures are actually injected into the physical system to observe their effects.

A designer would use either a FMEA or a FTA to identify the system's failure modes and their effects. For each effect the analysis will show what actions (if any) would prevent the effect from occurring, and if so, how much fault recovery time is allowed.

5.2.1.2 Fault Recovery Fault recovery methods try to fix the failures. Perhaps the most widely used method for hardware fault recovery is *redundancy* [46], where a faulty component is replaced with an identical, fully operational spare component. Redundant hardware has one distinct advantage: it can under normal circumstances completely restore the system's behavior to what it was before the failure occurred. Unfortunately, normal circumstances do not always exist, which can make redundancy impractical. For instance, in some systems mission essential equipment occupies all of the available space leaving no room for spare hardware. Another case is when system failure results from an unanticipated change in the operational environment. For example, high radiation environments can induce effects in metal-oxide-semiconductor devices [56] and high temperatures can lead to failures in voltage converters [70]. Simply replacing the original system with an identical copy accomplishes nothing if the environmental conditions still persist.

Another fault recovery method—and this is where EHW particularly excels—is *reconfiguration*. Reconfiguration methodically changes the failed system's design, both in component values and component interconnections, until a desired functionality is restored. Although there is no guarantee full functionality can be restored in all cases, reconfiguration is a viable and powerful fault recovery technique. In particular, reconfiguration can adapt existing hardware to work in unanticipated and potentially hostile environments—something redundant hardware can't do.

FDI methods must be predictable. Hence, FDI is inherently deterministic, which means EHW has nothing really to offer. Fault recovery, however, is an entirely different matter because it can be either deterministic or stochastic. Deterministic recovery methods are very effective whenever the precise nature of the failure is known. Unfortunately, in many (perhaps most) situations that would be presumptuous. A strictly deterministic recovery method, such as redundant hardware, is useless when faults are caused by operational environment changes since there is no reason to believe the new hardware won't also fail just like the old hardware failed. Redundant hardware may also be only marginally effective if the precise nature of the failure is unknown. For example, a collision with debris is a serious problem for orbiting satellites [47]. It will be hard to find out exactly how a satellite was damaged by a collision without conducting a firsthand visual inspection—something not always easy to do! Blindly replacing hardware without knowing what the problem is reduces fault recovery operations to outright luck. Under these circumstances stochastic recovery methods may be the only viable option primarily. Stochastic recovery has no predefined solutions so it is more flexible and adaptable than deterministic recovery. Stochastic recovery is therefore better able to deal with unforeseen situations.

Stochastic recovery methods use reconfigurable (i.e., reprogrammable) devices. These devices can actually be used for deterministic recovery by predefining a set of configurations, which can be downloaded whenever needed. This capability provides a flexible platform that can effectively deal with anticipated faults; reconfigurable hardware can adopt different functionality, which is important when there is limited room for spares. On the other hand, EHW combines the powerful search capability of evolutionary algorithms with the reconfigurable devices. Evolutionary algorithms conduct a stochastic search, which explains why EHW is considered a stochastic recovery method. EHW doesn't rely on predefined configurations—it evolves them. Evolution has the potential of creating unanticipated circuitry, which may be precisely what is needed to recovery from unanticipated faults.

It will be shown shortly that fault recovery is inherently a real-time process. This has strong implications not just for EHW, but any fault recovery method. Indeed, it is the one property that eliminates otherwise perfectly good recovery methods!

5.2.2 Real-Time Systems

This section provides a brief introduction to real-time systems. Interested readers are referred to some of the excellent books on this topic for further information (e.g., see [48]). We begin with a formal definition of a real-time system.

Definition: (*real-time system*)

Any system that is both logically and temporally correct.

Logical correctness means the system performs all of its assigned tasks and functions according to specifications without failure. Temporal correctness means the system is guaranteed (repeat, guaranteed) to perform these functions within explicit timeframes. fault tolerant systems qualify as real-time systems because FDI and fault recovery inherently have deadlines. In other words, a fault must be detected and isolated within a certain period of time after it occurs, and the fault must be corrected within a certain period of time after it is detected. Fault recovery may also have an expected start time. That is, regardless of how long it takes to do fault recovery, those fault recovery operations must commence within some certain time after the fault occurred.

The concept real-time is often misinterpreted. Some researchers (and most people in marketing departments) believe real-time means really, really fast. This interpretation is not correct. Real-time does not necessarily mean fast and fast does not necessarily mean real-time. Notice that the real-time system definition does not say that something has to be done really, really fast. In fact, temporal correctness does not require events to take place in milliseconds or even nanoseconds—it only requires it take place when it's needed. A simple example will illustrate these concepts.

Suppose a document must be sent from Chicago to London, and two delivery systems are available: surface mail with a guaranteed 3-day delivery time or e-mail with a guaranteed 5-minute delivery time. Note that both delivery systems guarantee delivery so logical correctness holds. Now we need to verify that temporal correctness holds. The e-mail delivery is orders of magnitude faster than surface mail, but that does not necessarily mean it qualifies as a real-time delivery system. It is the required delivery deadline that determines if temporal correctness is satisfied. For example, both systems are real-time systems if the deadline is six days because both systems are temporally correct. However, neither one is a real-time system if the deadline is three minutes because neither one is temporally correct.

Real-time systems are classified as hard or soft. These classifications indicate the consequences if temporal correctness requirements are not met. Failing to meet a temporal requirement in a soft system only causes degraded performance. On the other hand, similar failures in hard systems have catastrophic consequences—up to and including complete system destruction. The exact classification for fault tolerant systems depends on the nature of the faults and the consequences for failing to detect and correct them in a timely manner. Consider an autonomous deep-sea probe exploring the ocean bottom. Suppose any over-temperature condition can be tolerated for no more than five minutes or the probe self-destructs. Clearly fault recovery must be completed within five minutes to prevent impending loss of the probe. This qualifies as a hard system. However, suppose a fault only causes a minor loss of some sensor data after five minutes and this lost data could be recovered using extrapolation techniques. Fault recovery in this case could take considerably longer than five minutes without any dire consequences. This qualifies as a soft system.

5.3 CREATING FAULT TOLERANT SYSTEMS USING EHW

It was previously shown that EHW has emerged as a powerful hardware design method—which naturally suggests EHW methods could be equally useful for reconfiguring hardware in fault recovery operations. Either extrinsic or intrinsic evolution could be used. However, in many instances intrinsic evolution is necessary because the only real way to evaluate a configuration is to implement it and have it actually operate in the physical environment. *Intrinsic reconfiguration* refers to any search process that uses intrinsic evolution to find new hardware configurations.

Some recent work has shown EHW can be quite effective for reconfiguring existing hardware to overcome faults [58, 60, 65]. The ability of evolutionary algorithms to find good reconfigurations is not at issue here. Instead we are concerned with the issue of time. Most EHW-based studies rely on device simulators rather than physical hardware. This simulator use means the time to download a configuration, the time to program the device, and the time to conduct a fitness evaluation on the reconfigured hardware has largely been ignored—even though they dramatically affect the evolutionary algorithm's running time.

This notion of time is what differentiates EHW used for original design from EHW used for fault recovery. Original design has no real time limit (other than possibly annoyance from a long waiting time) because there are no real consequences for a long EA running time. Conversely, online systems—particularly safety-critical online systems—cannot operate indefinitely with faults. It is imperative that any faults be detected promptly and fault recovery operations commence before other things can go wrong. EA running time now becomes a factor in choosing the fault recovery method. *This suggests EHW, despite its power and flexibility, may not always be the best choice for a fault recovery method.* As we will see shortly, it all depends on whether or not the EHW approach can meet the recovery deadline.

It should be obvious that timing constraints must be considered whenever fault recovery—whether evolutionary algorithm based or not—is considered. Surprisingly, most EHW researchers who study fault tolerant systems focus almost exclusively on restoring functionality [52]. The EA running time is usually reported as either a maximum time or as an average time taken over many runs. The target application is usually not specified even though the application determines the fault recovery deadline. Average running time doesn't guarantee compliance with a hard recovery deadline. Even the maximum running time is worthless because there is no guarantee some future run won't take even longer.

Designers must contemplate a variety of factors before choosing a fault recovery method. In particular, fault recovery methods must remove the fault within a specified time frame. EHW can use either extrinsic or intrinsic reconfiguration, but we believe only intrinsic reconfiguration is of any use with online, deployed systems. We start by describing the advantages and limitations of intrinsic reconfiguration so designers can make an informed choice. In addition, several recommendations about the best ways to use intrinsic reconfiguration for fault recovery are included. No new methods to speed up the evolution are discussed. The reader should understand intrinsic reconfiguration properties are not application dependent and are not restricted to any particular type of evolutionary algorithm. Optimization issues are handled more appropriately within the context of a specific problem.

5.4 WHY INTRINSIC RECONFIGURATION FOR ONLINE SYSTEMS?

We are concentrating on autonomous fault tolerant systems—i.e., systems that can detect, isolate and ultimately recover from a fault without any human intervention. Whenever EHW is used for fault recovery, the evolution can be done either intrinsically with physical hardware or extrinsically with hardware simulations. Is one method better than the other? We believe for truly autonomous fault tolerant systems only the intrinsic method makes any sense for two reasons:

1. **Lack of In-Situ Computing Resources**

   Most EHW research is conducted in laboratories using Pentium PCs or UNIX workstations—resources not usually installed on a deep-space probe heading towards the planet Neptune! Many online autonomous systems operate in operational environments that don't provide much extra room. There simply isn't enough free space available for bulky computer hardware. There isn't room for redundant hardware either. Fortunately reconfiguring hardware provides a great deal of flexibility without consuming a lot of space. A more likely computing environment for reconfiguration operations in autonomous systems would be a single low-end microprocessor. Such a processor is relatively slow and can only address small amounts of memory. This increases simulation runtime and the accuracy of the hardware models, which negatively affects extrinsic reconfiguration methods. Conversely, intrinsic reconfiguration doesn't use hardware models, so there are no modeling inaccuracies. Low-end microprocessors are quite capable of running evolutionary algorithms used for EHW. In fact, compact genetic algorithms even run entirely on a single VLSI chip [51].

2. **Inherent Inaccuracies in Hardware Simulations**

   This limitation is best explained with an example. As mentioned previously, spacecraft can incur serious damage from collisions with orbital debris [47].

Unforeseen failures like these are hard to represent accurately in a simulation, and even then only if the precise nature of the failure is known. Unfortunately, getting accurate failure information is nearly impossible without firsthand observations—which is almost impossible to get from unmanned, autonomous spacecraft. This means any extrinsic reconfiguration starts out with a very severe problem: the simulation not only deviates from the true system behavior, but how much it deviates is unknown. Evolving a new hardware configuration under these circumstances is futile because there is no way to verify how the damaged system is really going to respond to that new configuration. On the other hand, intrinsic reconfiguration never has to make simplifying assumptions because the evolving hardware is directly interfaced to the damaged system. Hence, there is no doubt about how much functionality is restored in the reconfigured system.

The above two items should not imply intrinsic reconfiguration has no disadvantages. Fault recovery is inherently a real time process and it will shortly be shown that intrinsic reconfiguration can take surprisingly long. In fact, under certain circumstances it may take so long that it cannot be used for fault recovery! Our objective is first to show designers the limitations of intrinsic reconfiguration and second to provide the necessary tools for evaluating its potential as a fault recovery method.

5.5 QUANTIFYING INTRINSIC RECONFIGURATION TIME

In this section we derive a formula for estimating the intrinsic reconfiguration time and describe how the formula is used. The derivation assumes the hardware environment consists of a single processor (microprocessor or microcontroller), interfaced to a single reconfigurable device (FPGA, FPAA or FPTA). The processor is responsible for executing the evolutionary algorithm, converting configurations into a proper file format, downloading the configuration to the device, and running fitness tests to evaluate the configuration. The processor may or may not be dedicated to reconfiguration. That is, the processor may be engaged with other tasks whenever reconfiguration is not being performed.

The EA manipulates a genome that encodes all of the information needed to create a hardware configuration. During each generation λ offspring are created and each must undergo a fitness evaluation. There are three factors that contribute to the running time of a generation: (1) the time to program the reconfigurable device (t_{pgm}), (2) the time to conduct the fitness evaluation (t_{fit}), and (3) the overhead time of the EA (t_{oh}).

With intrinsic evolution an offspring's fitness evaluation cannot begin until the configuration information is physically downloaded and the device is programmed. The actual t_{pgm} value depends on the type of reconfigurable device. Table 5.1 shows the download time for several reconfigurable commercial-off-the-shelf (COTS)

TABLE 5.1. Programming Times for Various Popular Reconfigurable Devices

Device	Type	Size	Mfg	t_{pgm} (ms)
ispPAC10	FPAA	4	Lattice Semiconductor	100
AN220E04	FPAA	4	Anadigm	3.8
XC3020A	FPGA	64	Xilinx	1.5
Virtex XCV50	FPGA	1728	Xilinx	7
XC4085XL	FPGA	3136	Xilinx	192
APEX II EP2A70	FPGA	6720	Altera	12.5
JPL's FPTA2	FPTA	64	Fabricated by MOSIS	0.008

All are commercial COTS devices except the FPTA. The units for size are configurable logic blocks for FPGAs, modules for FPAAs and cells for the FPTA. Timing information on the FPGAs and FPAAs was extracted from vendor datasheets; timing information for the FPTA was furnished by Zebulum [71].

devices[1]. The reader is cautioned that programming times may be small, but they are not negligible; EAs often have populations with hundreds of individuals and they frequently run for thousands of generations.

In an intrinsic evolution t_{fit} is the test time. This test time accounts for the vast majority of the EHW run time. It is important to emphasize t_{fit} is application dependent because the test duration is determined by the scope of the fitness test.

t_{oh} covers the time it takes an EA to do all of the other tasks in each generation such as conducting binary tournaments or performing n-point crossover. In other words, this is the time it takes to do reproduction and selection each generation. The genome format associated with a particular reprogrammable device is always the same, so the overhead associated with running the evolutionary algorithm is roughly constant each generation regardless of the application.

The total intrinsic reconfiguration time for an EA run for k generations while procreating λ offspring per generation is

$$T_r(k, \lambda) = k\lambda(t_{pgm} + t_{fit}) + kt_{oh}. \quad (5.1)$$

Of course, T_r says nothing about the quality of the circuits produced after k generations. We will discuss reconfiguration quality at length in Section 5.6 and at that time show how it relates to T_r.

The first term represents the physical programming and fitness test time for all of the offspring created during the EA run. The second term accounts for the reproduction and selection operations conducted during the run. Both the number of generations (k) and the EA overhead time (t_{oh}) can dramatically affect reconfiguration time. k depends on the thoroughness of the search and how easily

---

[1]No attempt was made to promote one reprogrammable device over another. In fact, the devices described in the table were picked merely to show the broad range of what is available. There is, however, one caveat. Some reprogrammable devices, such as the ispPAC10, are EEPROM based. Hence, there is a limit to the number of times the device can be reconfigured. RAM based devices should be used if the population size and/or the number of generations is large.

the evolutionary algorithm escapes local optima. The overhead time depends on a variety of factors including how parents are chosen, the complexity of the reproduction operators, and so on. Additional time must be added to account for the time to convert each genotype into the proper file format needed for the download.

Some examples will illustrate how the formula is used. We have purposely set $t_{oh} = 0$ in these examples to focus on an enormously important issue: the surprising time it takes to program and evaluate reconfigurations[2].

Example 1:

> A generational GA intrinsically evolves an EP2A70 FPGA. The population size is 200 and 1000 generations are processed. From Table 5.1, $t_{pgm} = 12.5$ ms. Then the time spent just reprogramming the FPGA is $\lambda \cdot k t_{pgm}$ or about 42 minutes.

t_{fit} is responsible for the rather long search times often encountered in analog applications. For instance, suppose a proportional-derivative controller is implemented in an FPAA. The objective is to set the two (real-valued) constants inside the controller. First applying a step input and second by measuring the system's settling time determines the reconfiguration's fitness[3]. This fitness evaluation lasts at least as long as the settling time, which can be somewhat lengthy. Indeed, settling times of two minutes are not unheard of [63]. Under these circumstances, it wouldn't take a very large population size nor a large number of generations to make an intrinsic reconfiguration run for hours or even days before finishing.

Example 2:

> An AN220E04 FPAA is used to compensate for aging effects in a control system responsible for positioning a satellite's communications antenna. A generational GA run for 500 generations with a population size of 100 does the reconfiguration search. The system's step response is measured to determine if the compensation is correct. This step response test takes $t_{fit} = 625$ milliseconds to conduct. Hence, $\lambda = 100$, $k = 500$ and $t_{pgm} = 3.8$ ms. Substituting into the above Eq. (5.1) yields an intrinsic reconfiguration time of approximately 8.7 hours.

Reconfiguration times are meaningless until they are put into context. For instance, take Example 2 above. Suppose brief communication sessions with the satellite are scheduled at 10-hour intervals. A session may be skipped, but skipping two sessions in a row is not permitted. If a fault is detected just prior to a scheduled session, and if the error results in missing the session, then the fault recovery deadline is 10 hours. This is the worst-case scenario[4]. An almost

---

[2]These examples are only meant to show how long it can take to complete an intrinsic reconfiguration. The quality of that reconfiguration is ignored for the moment.
[3]Settling time is the time it takes for any oscillation to die out. Normally the system is considered to have settled if the output is within a few percent of its final value.
[4]Missing one session is permitted. If the fault were detected just after a scheduled session, the fault recovery deadline would be 20 hours.

9 hour reconfiguration time may seem quite long, but in this case it is perfectly acceptable because it is less than the mandatory 10 hour requirement. On the other hand, it would not be acceptable if communication sessions were scheduled at 6-hour intervals.

The only way to determine if there is a problem is to compare the reconfiguration time against the fault recovery deadline. This latter quantity is system dependent. No problem exists so long as the reconfiguration time is less than the recovery deadline. This time comparison adds a new perspective on intrinsic evolution and, at the same time, imposes a new requirement. Reconfiguration becomes a real-time process whenever it is used as a fault recovery method. Consequently, it is no longer sufficient to just talk about how an evolutionary algorithm is able to restore a circuit's functionality. These statements may show logical correctness, but without comparing the reconfiguration time against a deadline there is no proof of temporal correctness. Just reporting an algorithm's running time doesn't say anything about temporal correctness either. The key point is expressed by the following first principle:

First Principle of Fault Recovery

No fault recovery method can legitimately proclaim efficacy until it is proven to be both logically and temporally correct.

The validity of this principle is easy to see. If the recovery method isn't logically correct, then the problem can't be fixed. If it isn't temporally correct, then the problem can't be fixed soon enough to prevent other things from going wrong. Without proving logical *and* temporal correctness, there is no basis for claiming a fault recovery method is effective.

It is easy to prove if a fault recovery method is logically correct—try it and see if it fixes the problem. Proving temporal correctness, however, is more complicated because it really depends on conducting a thorough FMEA or FTA. This analysis should identify all potential faults and their effects on system performance [62]. One outcome of the failure analysis is the recovery deadlines. Temporal correctness is proven if a logically correct recovery is guaranteed to finish prior to the recovery deadline. Unfortunately, no EHW-based fault recovery method is temporally correct.

Comment: The reader is cautioned not to draw any premature conclusions from this last sentence. Just because EHW-based methods aren't temporally correct does not mean they are precluded from doing fault recovery. Indeed, in certain circumstances EHW may well be the only viable fault recovery method. This whole issue of EHW-based recovery and temporal correctness is an extremely important one, which is discussed at depth in the next section.

5.6 PUTTING THEORY INTO PRACTICE

We assume the system is fully designed and it meets all functional specifications. In a perfect world no fault recovery method would ever be used unless it is both logically and temporally correct. In principle FMET can prove logical correctness—inject the fault and see if intrinsic reconfiguration prevents the effect from occurring. However, proving temporal correctness for intrinsic reconfiguration is another story.

In reality temporal correctness cannot be *formally* proven for intrinsic reconfiguration because it uses an evolutionary algorithm. Since these algorithms are stochastic, independent runs won't always produce repeatable outcomes[5]. Hence, there is no way to guarantee a fixed number of generations will always find a desired circuit configuration.

It is essential that a proper perspective be kept when evaluating temporal correctness. Eq. (5.1) must be used in the correct way. A designer should not run the EHW algorithm and then compare it against the recovery time to verify temporal correctness. This is backwards. A designer should start with the recovery time, determined from an FEMA or FTA, and then use Eq. (5.1) to figure out how many generations are available to perform the intrinsic reconfiguration. In other words, don't run the EHW algorithm and then compare it against the recovery time. Instead, take the recovery time and then figure out how many generations can be processed. The EHW-based method has to be evaluated on what it can produce within this number of generations. If the final reconfiguration isn't good enough, and the reconfiguration can't be done in stages, then some other fault recovery method should be chosen.

What about convergence? Can we prove whether or not the EHW algorithm converges to a good solution? Unfortunately, the answer is no—and such a proof is not forthcoming any time soon! General global convergence in evolutionary algorithms has been explored (e.g., see [45, 50]). However, those theorems proved convergence in the limit—that is, without any restrictions on how many generations the algorithm was allowed to run. Any conclusions based on those theorems don't apply here because fault recovery has a deadline, which sets an upper limit on the number of generations an algorithm can run. Bear in mind the evolutionary algorithm is a non-deterministic algorithm. Hence, any convergence proof for an EHW-based algorithm would have to, in effect, guarantee a stochastic algorithm converges to an optimal solution in a fixed number of time steps. This is a futile endeavor.

Yet even if temporal correctness cannot be proven, this does not mean intrinsic reconfiguration can't be used for fault recovery. It is also a reality that no system is perfect, always capable of performing when called upon. Systems do fail, so all recovery methods—whether or not they use evolutionary algorithms—have a finite probability of not executing. Consequently, the primary concern is to

---

[5]By the way, all decisions about temporal correctness should be made on maximum algorithm running times and not average running times.

reduce the likelihood of failure. All a designer can do is minimize the failure risk to an acceptable level. There are ways to minimize risk when using intrinsic reconfiguration, but that depends on whether the fault is anticipated or unanticipated. We will shortly see that in some circumstances the only way to deal with some faults is to modify the definition of logical correctness.

5.6.1 Minimizing Risk With Anticipated Faults

Anticipated faults are those faults identified while conducting an FMEA or FTA. These faults are precisely known and their effects are observed. Moreover, the fault recovery time is known because the analysis tells the time delay between the fault and its effect.

One nice feature of anticipated faults is there is no need to evolve anything *in-situ*. A logically correct configuration found ahead of time can be stored in memory prior to deploying the system. Whenever the fault occurs, simply download the prestored configuration. Under these circumstances temporal correctness is trivially achieved because the reprogramming time is quite small (see Table 5.1).

It was stated above that a FMET would verify logical correctness. But the first thing—before even conducting the FMET—is to see if intrinsic reconfiguration has any chance of meeting a recovery deadline. Here simulation can be useful[6]. Once the fault is identified a designer can envision the type of circuitry needed for fault correction. The designer could construct an evolutionary algorithm to search for hardware configurations and the simulator can evaluate them. It isn't necessary to find the exact desired configuration because the goal is only to approximate the evolutionary algorithm running time. Hence, the search can terminate when a sufficiently close enough configuration is found. Equation (5.1) provides an estimate of the physical hardware reconfiguration time which can then be compared against the fault recovery deadline. This procedure will quickly show if intrinsic reconfiguration is feasible.

The FMET should be conducted if and only if the reconfiguration time is less than the recovery deadline. The best way to minimize any risk is to allow sufficient slack time. In other words, the intrinsic reconfiguration should end well short of the recovery deadline. The definition of "well short" of course depends on the system and the consequences of missing the recovery deadline. In safety-critical systems—that is, systems where failure can lead to total system destruction, physical injury or even death—the recovery deadline should be several times longer than the reconfiguration time. That requirement can be relaxed for systems that aren't safety-critical, but reducing the reconfiguration time will always decrease the risk. *This means the designer must know what constitutes an acceptable level of risk before specifying the maximum reconfiguration time.*

Another way of reducing risk uses concepts taken from on-board preventive maintenance programs [69]. The main idea is to do fault recovery in stages. In

---

[6]Some COTS device vendors provide simulators for their products. Custom FPAAs and FPTAs and some mixed-signal devices can use PSpice for simulation.

other words, during each stage, which is a fixed time period, intrinsic reconfiguration is only expected to find a somewhat improved solution. The evolutionary algorithm uses the final population from the previous stage as the initial population in the current stage. The (still faulty) system has degraded performance, but it remains online between stages. This process repeats until full recovery is achieved.

An example will help to explain this multi-stage recovery process. Suppose a fault in an amplifier circuit not only reduces the gain, but it also causes an increased load on the power supply. This increased load creates a heat build-up which, if not eliminated within 45 minutes, will generate even more faults. An intrinsic reconfiguration cannot finish within 45 minutes, so instead it is broken into two 30 minute stages. In the first stage the only objective of reconfiguration is to reduce this excessive power supply load. Once the heat has dissipated, a second recovery stage can reconfigure the amplifier to restore the gain. Notice how the reconfiguration is prioritized.

No fault recovery procedure can remove all risk. This means other actions may be necessary. See MIL-STD-882D for some additional risk mitigation measures [61].

Redundancy plays an important role in fault tolerance and it can be especially useful against anticipated faults. The idea is to provide multiple (usually three) copies of a circuit and let a majority vote decide the output. A fault that occurs in just one of the redundant circuits is always detectable. But faults in two or more redundant circuits will only be detectable if they affect different bits in the output. Consequently, designers try and make the circuit copies diverse. The main idea here is this: instead of evolving a single circuit the designer evolves an ensemble of circuits and lets a majority vote decide the output. Each circuit in the ensemble would be intrinsically evolved during a FMET.

However, just evolving a set of different (albeit, logically correct) circuits may not yield the best ensemble. Each faulty circuit generates an error pattern. The best ensemble would have little or no error correlation between the circuits so the same fault in different circuits would tend to create different outputs—which makes the faults detectable. So what is the best way to create such an ensemble? Some recent work by Schnier and Yao [64] suggests an answer. During the EA run they calculated an error pattern vector which indicates how well a circuit performs both with and without injected faults. A scalar product of two error pattern vectors gives the cross-correlation between two circuits.

The fitness computation of a candidate circuit was based on the average correlation between the candidate and all other circuits in the population. This rewards circuits that

- have fewer errors without faults;
- have fewer errors with faults;
- have an error distribution that differs from that of other circuits in the current population.

This latter item is particularly important because it allows the EA to find circuits with a lower fault sensitivity. The best circuits to pick for the ensemble are the ones with the lowest cross-correlation.

5.6.2 Minimizing Risk With Unanticipated Faults

Unanticipated faults are caused by events beyond a designer's control. Two sources of unanticipated faults are persistent, unexpected environmental changes or damage resulting from exogenous events such as collision with orbital debris. In many instances the exact change in the environment or the extent of damage cannot be accurately determined. Consequently, unanticipated faults frequently have unpredictable effects. Without knowing the effect, it is impossible to state the recovery time; temporal correctness is no longer relevant. However, logical correctness may still be important but that depends on the source of the unanticipated fault. In some cases the definition of logic correctness won't change, while in other cases the definition will change.

It must be emphasized that "unanticipated" does not necessarily mean "undefined". Unanticipated can also mean "unplanned for". This is an important distinction which affects how logical correctness is redefined. But before we can discuss that, it is important to fully understand the difference between unplanned for events and undefined events.

A designer may be fully aware of a failure mode and may have even thoroughly studied its effects. Everything about this failure and its possible effects could be known. Nevertheless, the designer may honestly believe the failure can be safely ignored because there is such a low probability of ever seeing it. Consequently, no plans are made to either prevent the fault or to deal with its effects. In particular, no special precautions are built into the design, and no recovery mechanism to handle it is in place. There is no need to worry about it because there is no anticipation it will occur.

The other possibility is the failure mode was inconceivable from the very beginning. There are two reasons why: either the designer was not aware that a circuit could fail in a particular way, or there is no way to even guess about the fault ahead of time. Fortunately there is a way to help forestall the first reason. One of the advantages of conducting a FTA is the failure effect is defined first and then one hunts for the cause. In some cases an FTA can identify failure modes a designer never imagined. But the second reason is more difficult to contend with. These faults often result from exogenous events the designer has no control over. A satellite colliding with orbital debris is a good example [47]. There is no possible way a designer could know beforehand what type and extent of damage results from a collision—i.e., the fault and its effect can't be defined. How then, could a designer safeguard a circuit or restore its operation after such an event? Besides, there is no way to adequately test any precautionary methods or recovery techniques that might be conceived.

It should come as no surprise that the same fault could result from an anticipated event or an unanticipated event. Suppose a space probe is expected to encounter a high radiation environment. Certain precautions could be incorporated into the design to prevent any problems. For example, rad-hard components could be used and the enclosure could have special shielding installed. In this instance the high radiation environment would be an anticipated event. However, if the designer believes the chances are slim of encountering a high radiation environment, then these precautions might not be included perhaps to reduce costs. In other words, the high radiation environment is not planned for in the design. This means if the high radiation environment is subsequently encountered, it would qualify as an unanticipated event.

How (and if) the definition of logical correctness must change depends on whether or not the event is anticipated and what capability the recovery method can provide. This is illustrated by two categories of events: a changed environment (anticipated event) or an exogenous act (unanticipated event).

1. **Changed Environment**

    Component characteristics can change when the operational environment changes [56, 70]. A new hardware configuration might exploit those changed characteristics to restore any lost functions. For example, Keymeulen et al. [58] successfully reconfigured a FPTA to restore functions lost due to temperature increases. Logical correctness always holds for these types of problems because the recovery objective is the same as it is for anticipated faults. That is, the goal is to completely restore all original system functions.

2. **Exogenous Acts**

    Exogenous events create havoc with an operating system because there is no way to predict what faults will be induced. There is likewise no way to predict the ultimate effects. Consequently, logical correctness may be unattainable because there is no guarantee the available fault recovery method can deal with an unspecified failure mode. A decision must be made about the potential effectiveness of a given recovery method before activating it. Intrinsic reconfiguration done *in-situ* does have one advantage in this regard: in principle it can produce circuit figurations never imagined by the designers.

In truly catastrophic situations there may be no hope of restoring any loss functionality to any significant extent. Under these circumstances the corrective action switches from restoring system functions to guaranteeing system survival—which contradicts the normal interpretation of logical correctness in a fault tolerant system. Intrinsic reconfiguration would initially just search for a stable system configuration that stops any fault migration. For example, compensators can keep a system stabilizable even when a sensor or actuator completely fails [67]. Any other recovery actions may have to be done after the damaged system is taken offline.

One final aspect of unanticipated faults is worth mentioning. As noted above temporal correctness is ignored under certain circumstances. Does this mean the first principle of fault recovery is invalid? Not at all. *The first principle always holds when recovery is from an anticipated fault.* The reason is an anticipated fault can be thoroughly analyzed to determine its effect and recovery time. Temporal correctness requirements are therefore clearly defined. Conversely, an unanticipated fault is never analyzed, so any discussion about recovery time is meaningless. Temporal correctness cannot be defined in these cases.

5.6.3 Suggested Practices

Greenwood et al. [52] suggested evolutionary algorithms designed for reconfiguration searches perform best if they have high selection pressure and if they emphasize mutation for reproduction. In principle, any type of evolutionary algorithm could be used for a reconfiguration search, but from a practical standpoint genetic programming algorithms should be avoided. Genetic programming algorithms designed for EHW problems are put on large multiprocessor systems to abridge their long running time [57, 68]. This computational requirement is especially problematic for fault tolerant systems because, if there isn't enough room for redundant hardware, then there isn't enough room for a large multiprocessor system either. It seems unlikely a full-fledged genetic programming search, run on a single processor, could finish quickly enough to meet a fault recovery deadline of only a few hours.

Some devices can be partially reconfigured—that is, configuration changes can be restricted to only small portions of the device. The idea is to identify the particular region in the bit stream that requires reconfiguration. A header is then wrapped around this bit stream data to identify the address at which to start reconfiguration [55]. Partial reconfiguration reduces t_{pgm}. This reduction may be important in situations where t_{pgm} and t_{fit} are of the same order of magnitude.

5.7 EXAMPLES OF EHW-BASED FAULT RECOVERY

This section contains several examples to illustrate the ideas introduced in the previous section. Each example is taken from published work. Important lessons for EHW designers are at the end of each example. Not all of the publications report the EA running time making it impossible to say anything about temporal correctness—in apparent violation of the first principle of fault recovery! Nevertheless, they are instructive because at least they show EHW methods can achieve logical correctness.

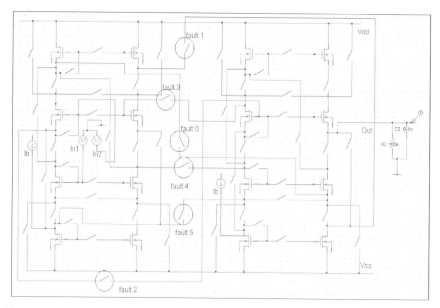

Figure 5.2. Schematic showing how two cascaded FPTA cells are used to construct a fault tolerant analog multiplier. The two cells are interconnected by 6 switches. These interconnection switches, denoted by fault 0, fault 1, and so on are used to inject faults. See Chapter 3 for details on the FPTA cells.

5.7.1 Population vs. Fitness-Based Designs

Keymeulen et al. [58] used a genetic algorithm to evolve fault tolerant circuits in an FPTA[7]. Two different approaches for evolving fault tolerant circuits were proposed:

1. *Fitness-based fault tolerant design*: Pre-defined faults are injected into the population during the genetic algorithm run. The underlying philosophy is evolving a circuit in the presence of a fault should promote the design of a circuit that can tolerate that fault.
2. *Population-based fault tolerant design*: A population of circuits is evolved without any injected faults. A fault is then injected and the best performing individual is extracted. If the best individual's performance isn't good enough, the genetic algorithm is restarted, with the fault injected, using the last available population as the initial population. This process "repairs" the circuit.

The experiment was to evolve a fault tolerant analog multiplier using an FPTA. Figure 5.2 shows how two cells of the FPTA, interconnected by 6 switches, form

---

[7]Both an analog and a digital design were synthesized, but only the analog experiment is described here.

the multiplier circuit. These interconnection switches could be opened or closed to inject faults into the circuit. A 54-bit binary string forms the genome: 2 sets of 24 bits controlling the internal cell switches plus 6 bits for the interconnection switches.

In1 and In2 are the analog multiplier's inputs. These DC inputs range from 1 to 4 volts in 0.3 volt increments. For any particular inputs In1(i) and In2(j), the measured output value is Out(i, j) and the target output value is In1(i) · In2(j). The fitness of a configuration C is given by

$$\text{fitness}(C) = 1 - \sqrt{\frac{\sum_i \sum_j [\text{In1}(i) \cdot \text{In2}(j) - \text{Out}(i, j)]^2}{n}},$$

where n is the number of samples. Notice the fitness increases as the RMS error to the target decreases.

All configurations were extrinsically evolved using a SPICE circuit simulator. The experiments used a genetic algorithm with

population size: 128
genome: 54-bit binary string
uniform crossover probability: 0.7
mutation probability: 0.04
tournament: binary
generation gap: 50%

1. *Population-Based Experiment*: The genetic algorithm converged after 60 generations. (No faults were injected yet.) The best fit individual was subjected to one of the following faults (see Figure 5.2):

    faults 0, 1, 4 are opened switches
    faults 2, 3, 5 are closed switches

    These faults only degraded the performance of the best fit individual—i.e., no fault caused a total circuit failure. Other individuals in the same population (called "mutants") were also checked to see how they perform with an injected fault. Fault 2 did not affect the best fit individual. Faults 0, 1 and 4 degraded the best fit individual to an unacceptable level, and no mutant exhibited acceptable performance either. However, after restarting the genetic algorithm, with the fault injected, a repaired individual with quite acceptable performance was found usually in less than 100 generations. This clearly demonstrates repair is an effective fault recovery mechanism. Table 5.2 shows all of the results. Notice fault 3 and 5 did not require restarting the genetic algorithm because a good performing mutant was already in the population.

2. *Fitness-Based Experiment*: The same 6 faults defined above were used in this experiment. However, each individual was evaluated 7 times: once with no faults and then once with 1 of the 6 faults injected. An individual's

TABLE 5.2. Fitness Results for the Population-based Experiment

	Best Individual	Best Mutant	Self Repaired
no fault	0.9229	0.9229	
fault 0	0.3797	0.4141	0.8833
fault 1	0.4169	0.6074	0.8664
fault 2	0.9229	0.9229	
fault 3	0.4386	0.9168	
fault 4	0.6674	0.8032	0.9168
fault 5	0.6002	0.9147	

TABLE 5.3. Fitness Results for the Fitness-based Experiment

	Best Individuals	Self Mutants
no fault	0.9082	0.9082
fault 0	0.8592	0.8758
fault 1	0.9082	0.9082
fault 2	0.8459	0.8877
fault 3	0.9052	0.9052
fault 4	0.8697	0.8807
fault 5	0.9082	0.9082

fitness was the average of the 7 evaluations. After 50 generations the genetic algorithm converged to the best fit individual. Table 5.3 summarizes the results.

The population-based approach provided slightly better performance for most of the faults and with less computational effort. But that should not imply a fitness-based approach has no advantage. Indeed, the two approaches have very definite benefits when used in the proper situations. A population-based approach is perfectly suited for handling unanticipated faults—something the fitness-based approach cannot do as well because it evolves circuits evaluated against a set of anticipated faults and it does not restart the evolutionary algorithm. But in other circumstances those restrictions are not necessarily disadvantageous.

First of all a fitness-based approach evolves individuals that are subjected to multiple faults. Granted, those are anticipated faults, but the fitness-based evolution is more likely to produce a single circuit robust to those faults. Second, the evolution can be done before the system is put into its final operational environment and the best performing individuals can be stored for later use.

Now evolution is no longer required when an anticipated fault is detected; fault recovery consists of a single download of the most appropriate prestored configuration. This makes fitness-based approaches ideal for systems with extremely short recovery deadlines because it combines the power and flexibility of an evolutionary search with the appeal of a minimal recovery time.

5.7.2 EHW Compensators

Greenwood et al. [53] studied fault recovery for a class of systems without access to its interior for repair or replacement of any failed components. In other words, the systems are treated as "black boxes" that could not be fixed because there is no way to get to the failed circuitry inside. The only way to restore service—even degraded service—is by inserting a compensation network into the control loop. Figure 5.3 shows a block diagram. Even though a closed-loop system is shown, it could just as well be an open-loop system. The compensator is implemented in a reconfigurable analog device (FPAA) whose configuration is intrinsically evolved.

The plant is a linear continuous system whose dynamics are governed by a constant coefficient differential equation. All plant failures are manifested by a sudden change in the plant's bandwidth. These faults could result from a sudden component failure or more gradually if due to aging effects. For example, consider a simple RC low-pass filter. The pole would move to infinity if the capacitor suddenly opens, which increases the bandwidth. However, not all bandwidth changes are abrupt. The filter bandwidth would gradually change over time if the capacitor value slowly drifted due to aging effects. In real-world systems any bandwidth change alters the impulse response and could, in the worst case, cause the system to fail.

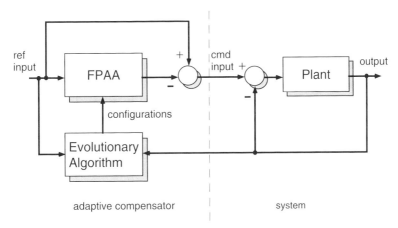

Figure 5.3. Environment for intrinsic evolution testing. The system is a black box without any access to its internal circuitry. The reference input is any test input signal used to evaluate the FPAA configurations created by the evolutionary algorithm.

The investigated plants are a 3rd-order system, which reasonably approximates a higher order plant's behavior. The plant transfer functions are

$$G_p(s) = \frac{K_p}{\Pi_{i=1}^{3}(s+p_i)}, \qquad (5.2)$$

where K_p is a positive gain, p_1 is a real pole and p_2, p_3 are a pair of complex conjugate poles. Aging effects do not alter K_p but they do alter the plant's pole locations, which changes the plant's bandwidth. Any bandwidth reduction makes the plant's response more sluggish, whereas any bandwidth increase could make the plant susceptible to high frequency noise.

All plant transfer functions were implemented with operational amplifier circuits. Different resistor values could be switched into the circuit which allowed an easy change of the transfer function to simulate errors. Figure 5.4 shows the 3rd-order plant circuitry.

The purpose of the compensator network is to restore, as much as possible, the plant's bandwidth to its original value while keeping the overall gain the same. The compensator transfer function is

$$G_c(s) = K_c \frac{(s+a)}{(s+b)}, \qquad (5.3)$$

where K_c is a positive gain, $a/b < 1$ for a lead compensator and $a/b > 1$ for a lag compensator. A lead (lag) compensator increases (decreases) the uncompensated plant's bandwidth. First consider the lead compensator. With $a/b < 1$ the zero is closer to the origin of the S-plane than is the pole. A bode plot of this transfer functions shows the magnitude increases at low frequencies because of the zero

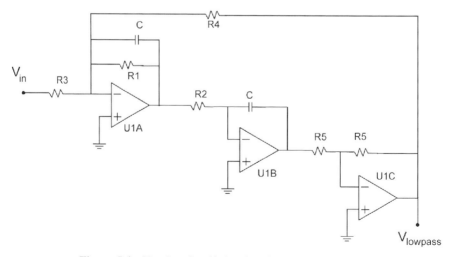

Figure 5.4. Circuitry for third-order plant transfer function.

and then levels off at a higher frequency because the pole cancels the zero's effect. This increases the compensated plant bandwidth. A lag compensator has the opposite effect because $a/b > 1$ places the pole near to the origin of the S-plane.

The programmable analog device used in this investigation is the Lattice Semiconductor ispPAC10. (See Chapter 3 for a detailed description of this device.) Unfortunately, there is a problem. The transfer function shown in Eq. (5.3) is minimally implemented with two resistors, two capacitors and one operational amplifier—a circuit that cannot be created in a ispPAC10 because the required component interconnections can't be programmed. Two points are important to make here: (1) The ispPAC10 is not the only reconfigurable device that is incapable of implementing a minimal circuit configuration. Indeed, this limitation is common to all COTS FPAAs. (This issue is discussed shortly in more detail.) and (2) Sometimes a little thought may find a solution despite device constraints.

One way of creating a compensator transfer function is to add together the outputs of a high-pass filter and a low-pass filter. Consider the circuit in Figure 5.5. This circuit yields a high-pass filter transfer function $H(s) = s/(s+a)$ and a low-pass filter transfer function $L(s) = a/(s+a)$. Notice that the pole location for both filters is at $s = -a$, which is set by the capacitor used in the integrator. Now, with positive gains K_1 and K_2, form the transfer function

$$G(s) = K_2 H(s) + K_1 L(s)$$
$$= \frac{K_2 s}{(s+a)} + \frac{K_1 a}{(s+a)} \qquad (5.4)$$
$$= K_2 \frac{(s + \zeta a)}{(s+a)}; \zeta = \frac{K_1}{K_2}.$$

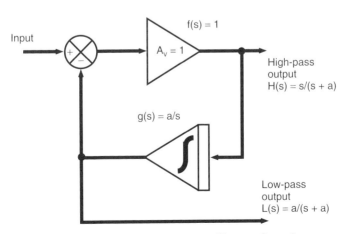

Figure 5.5. High-pass and low-pass filter configurations.

A lead compensator has $K_1 < K_2$ whereas a lag compensator has $K_1 > K_2$. Fortunately, it is very straightforward to create such a circuit in the ispPAC10 using three analog blocks. This configuration is shown in Figure 5.6.

Two sets of experiments were run. The first set of experiments investigated fault recovery while the second set investigated aging effects. In both sets the system was assumed to have undergone a change in bandwidth, and the reconfigurable analog device had to evolve a compensator to help restore the bandwidth. For convenience, the plants were referred to as original, reduced bandwidth, and increased bandwidth. The original system was fault-free whereas the other two systems were faulty. The objective was to evolve a compensator that restored the bandwidth of the faulty systems to that of the fault-free system. The plant transfer function was

$$G_p(s) = \frac{K}{s^3 + a_2 s^2 + a_1 s + K}. \tag{5.5}$$

Table 5.4 gives the coefficient values and the -3 dB points of the three plants investigated. These 3rd-order systems have a low-pass characteristic. The plant

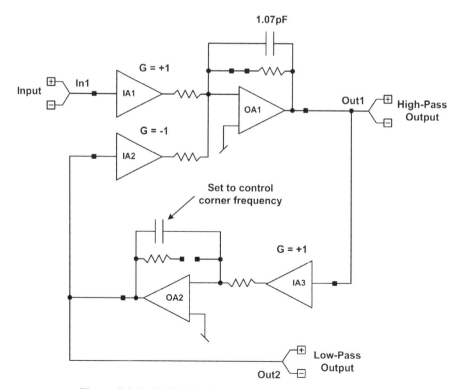

Figure 5.6. ispPAC10 implementation of a lead compensator.

TABLE 5.4. Transfer Functions Coefficients Used in Eq. 5.5 for the Original Bandwidth

BW	−3 dB point	K	a_2	a_1
original	44.25 KHz	1.097×10^{16}	3.422×10^5	7.605×10^{10}
reduced	31.07 KHz	5.597×10^{15}	3.422×10^5	5.185×10^{10}
increased	53.00 KHz	1.646×10^{16}	3.422×10^5	1.007×10^{11}

The reduced bandwidth and the increased bandwidth plants. The original plant is fault free. The reduced (increased) bandwidth plant requires a lead (lag) compensator. The original system is uncompensated.

bandwidth was modified by changing a single resistor value in the circuit shown in Figure 5.4.

The evolutionary algorithm evolves a population of 20 configurations over 200 generations. Both recombination and mutation are used, but recombination is only applied with probability 0.4. These reproduction operators select the amplifier gains and a capacitor value to set the frequency characteristics. (This determines the K_1, K_2 and a values in Eq. (5.5).) However, it turned out that it is necessary to also evolve one small portion of the circuit structure. It is not possible in the ispPAC10 to choose a gain of 0 for amplifiers IA1 or IA2 (cf. Figure 4.24). The only way to get zero gain is to remove the amplifier. Fortunately, the interconnection network in the ispPAC10 internal architecture permits amplifier removal. Therefore, two switch positions are added to the genome.

The same fitness function is used regardless of which compensator is evolved. The fitness of configuration C is given by

$$\text{fitness}(C) = \frac{1}{\sum_{i=1}^{5}[M(i) - M^*(i)]^2},$$

where $M(i)$ is the compensated system's magnitude and $M^*(i)$ is the magnitude of the original (error-free) system at frequency i. The five frequencies were chosen with two of them in the passband, two of them in the stopband, and one at the −3 dB point, which was 44.25 kHz. It may appear as if the fitness could be determined by just checking where the −3 dB point of the compensated system is; the closer to 44.25 kHz the higher the fitness. Unfortunately, such an obvious fitness function could produce the trivial solution: simply increase or decrease the open-loop gain. The goal is to make the passband and stopband gains as close as possible to that of the original system. Consequently, samples from the passband and stopband are included in the fitness function.

A large number of runs were used to verify that the technique could repeatedly evolve an acceptable compensator. Figure 5.7 shows the bode plots for the faulty increased bandwidth system and the restored (compensated) system from a typical run. (The original system bode plot is also added for reference.) Figure 5.8 shows bode plots for the reduced bandwidth systems. A decent compensator could usually be evolved in approximately 50 generations.

122 PUTTING EVOLVABLE HARDWARE TO USE

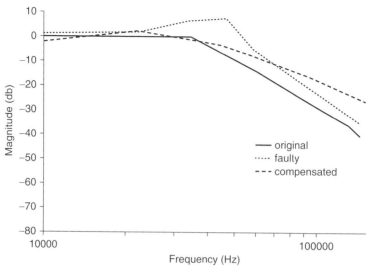

Figure 5.7. Bode plot for the increased bandwidth faulty plant experiments. Changing a resistor value in the plant caused the bandwidth shift, which emulates a failure. Note the faulty plant is somewhat underdamped. The compensated frequency response is produced by cascading the faulty plant with an evolved lag compensator.

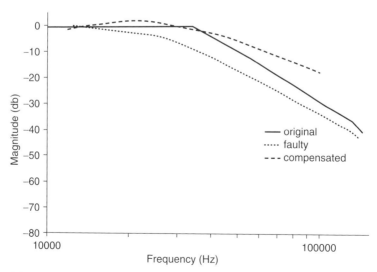

Figure 5.8. Bode plots for the decreased bandwidth faulty plant experiments. Changing a resistor value shifts the bandwidth, which emulates a failure. The compensated frequency response is produced by cascading the faulty plant with an evolved lead compensator.

The stopband characteristics of the lead compensated system are not as good as they should be. Nevertheless, the passband characteristics and the bandwidth in both cases are quite good. The authors conjecture the stopband characteristics would be better if more stopband samples were taken. Unfortunately intrinsic evolution of analog circuits is notoriously time consuming [54] and taking more samples extends that time even further. Each ispPAC10 configuration took about 5-6 seconds to evaluate with magnitudes sampled at five frequencies. Five samples seem to be a reasonable tradeoff between solution quality and an acceptable fault recovery time.

A second set of experiments investigated aging effects. Aging scenarios are different because the changes are more gradual and monotonic. For example, prolonged exposure to low temperatures can cause a slight decrease in capacitor values, which would tend to move the plant poles away from the S-plane origin—that is, the bandwidth increases.

Once again the 3rd-order system with a −3 dB point at 44.25 KHz was used as the fault-free system. A fault is induced into the plant by decreasing the value of resistor R1 (cf. Figure 5.4), which increases the bandwidth. After this resistor value change, which moves the −3 dB point to around 51 KHz, a lag compensator was evolved which successfully restored the bandwidth to near its original value. R1 was decreased again and this time the −3 dB point of the compensated system moved to about 57 KHz. An attempt to re-evolve the compensator was only partially successful. These results are depicted in Figure 5.9.

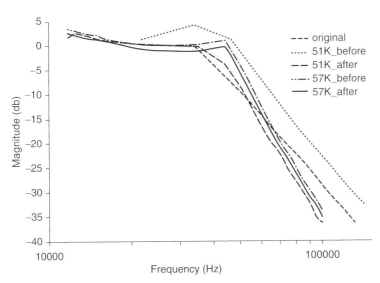

Figure 5.9. Bode plots for the aging experiments. 51 K-before and 51 K-after refer to the uncompensated and compensated plant response after the first resistor value change (see text). 57 K-before and 57 K-after shows the response after the second resistor change.

There is a simple explanation for why the results are only partially successful. The ispPAC10 provides 128 choices for each capacitor with the maximum capacitor value slightly below 62 pf. The authors point out that the best configuration compensator in the aging experiments had this value, which suggests a larger value would have been able to provide better compensation. This raises an important point about using reconfiguration as a fault recovery method. It is imperative that the reconfigurable analog device provides the component values needed to work over the frequency ranges of interest. Note that the 57 KHz −3 dB point was for the compensated system. Since a lag compensator was already installed, the original uncompensated plant −3 dB point would have been much higher with the second resistor value change. Clearly the ispPAC10 is not suitable for this range because the available capacitors are not large enough.

One problem not anticipated is the limitations naturally imposed by an FPAA architecture. A solution sometimes requires some clever tricks. For example, each ispPAC10 analog block contains resistors, capacitors, op amps and a programmable interconnection network. This does not, however, mean all possible circuit configurations using these components can be made. A basic op amp inverting amplifier only requires two resistors. Look at the feedback network in the op amp at the top of Figure 5.6. Although the resistor can be removed by programming the switch to be open, the capacitor cannot be removed—which makes it impossible to make a basic inverting amplifier. Nevertheless, if one chooses the smallest available capacitor value (1.02 pf), the single pole is located at a very high frequency. This is a compromise configuration, but it does approximate the performance of the basic op amp inverting amplifier at low frequencies.

We are not suggesting only the ispPAC10 is susceptible to architectural constraints. Just the opposite is true because all COTS reconfigurable analog devices at the same level of circuit granularity suffer from similar limitations. This situation is the norm and for good reason since the number of ways of connecting a set of components is surprisingly large. For instance, consider a resistor and an op amp. The resistor can pull-up, pull-down, or be in series with either op amp input or the op amp output. It could also connect the two inputs or connect either input to the output. Hence, there are twelve possible circuit configurations with these two components. Real-world analog circuits are far more complex than this, which means the search space for circuit topologies is enormous—and we haven't even considered choosing component values!

This situation has important implications for analog EHW researchers. We conjecture evolving circuit configurations is probably impractical at the system level, but will be okay for simple analog circuits such as oscillators or comparators. Fine granularity devices like the field programmable transistor array [66] are perfect for these latter type of applications. But for more sophisticated analog circuitry, such as control system compensators, it is probably best to fix the configuration and only evolve the component values because this can greatly reduce the computation effort. In part this is because the designer may be forced to use non-intuitive circuit configurations to work around FPAA architectural constraints. An example was shown in Figure 5.5 where a high-pass and a low-pass filter were combined to implement a compensator in the ispPAC10. It seems unlikely that an evolutionary algorithm could "figure out" this novel circuit configuration within a reasonable timeframe. However, evolutionary algorithms are ideally suited for choosing good component values in complex analog circuit designs.

5.7.3 Robot Control

In Tyrrell et al. [39], a novel evolutionary algorithm is developed that is based only on a fitness dependent mutation rate. One major objective of this work was to produce an evolutionary system that was implemented completely on COTS hardware. As an example of the new system, an FPGA-based controller for a mobile robot is presented. The controller is made up of LUT (look up tables) which perform the mapping from sensor data to actuator (commands to the motors) and it is evolved using the effective evolutionary algorithm developed. The experimental results on a Khepera robot show that the method can successfully evolve a robot controller for autonomous navigation to avoid collision in an unknown or dynamically changing environment even if sensor faults occur prior to evolution or after a successful member of a population has be evolved.

An excellent summary of work conducted in the general area of evolutionary robotics can be found in Nolfi and Floreano's book [29]. Recent work has used evolutionary processes to evolve the structure of neural controllers for robotic applications. In [14] the choice of a fitness function for the evolution of autonomous robotics that must navigate in an open-environment by avoiding obstacles was studied. In [15], an incremental evolutionary approach was applied to a particular set of robotic tasks. In the first stage a controller was evolved to avoid obstacles, in the second stage, a controller was evolved that approaches a target object by avoiding obstacles in the environment. The first stage was used as the initial population for stage two of the evolution. Both papers reported good results. However, the current experimental setup used a SUN SPARCstation 2 for all evolutionary processes and would therefore be considered extrinsic.

Genetic Algorithms have been applied to EHW because of its binary representation, which matches perfectly with the configuration bits used in FPGAs. A number of papers have been published to evolve on-line FPGA-based robot controllers using genetic algorithms [13, 16, 34, 35] using COTS. One of the main problems of evolving on a FPGA is the Genotype-Phenotype mapping. In the work reported in this section, an effective method to solve this problem using intrinsic EHW is proposed. The problem of adaptation of autonomous robot navigation in changing environments consists of finding a suitable *function F* (the controller) which maps the inputs from sensors to the outputs (control signals to the motors).

In many evolutionary algorithms, whether intrinsic or extrinsic, once the termination criterion has been met the evolutionary process stops and the best individual within the population is used in the implementation. An alternative approach, and one taken in this example, is to allow continuous evolution throughout the lifetime of a system. The evolutionary process continues as described above, however, once a member has been chosen for implementation the evolutionary process does not stop. Such a continuous process allows a system to be more responsive to environmental changes. In this example it is shown that evolution can cope with errors during runtime. The fitness might drop at the instant the error is activated, but the evolutionary process autonomously deals with this drop in fitness and recovers to an acceptable fitness level. Hence, an acceptable level of functionality is acquired over a number of generations.

The Virtex family [43] uses a standard FPGA architecture as in Figure 5.10. The logic is divided into an $N \times M$ structure of configurable logic blocks (CLB), each block contains a routing matrix and two slices. Each slice contains two lookup tables (LUTs) and two registers. The inputs to the slice are controlled through the routing matrix which can connect the CLB to the neighboring CLBs through single lines (of which there are 24 in each direction) and hex lines which terminate six CLBs away. The CLB configuration and all routing is done using the JBits classes described below.

It is possible to route two outputs together creating a bitstream which will damage the device. Consequently, only the logic functions of the robot controller will be evolved instead of both the logic and the routing. That way only safe controller will be evolved in these experiments.

The JBits [44] used for programming the FPGA are a set of Java classes which provide an Application Program Interface (API) into the Xilinx Virtex FPGA family bitstream. This interface operates on either bitstreams generated by Xilinx design tools or on bitstreams read back from the physical hardware device itself. Furthermore, this interface provides the capability to not only design but to dynamically modify circuitry inside the Xilinx Virtex series FPGA devices. The programming model used by JBits is a two dimensional array of CLBs, where

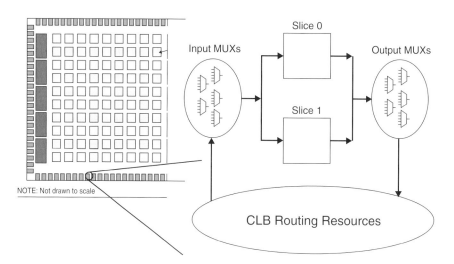

Figure 5.10. Virtex architecture [43].

each CLB is referenced by a row and column. All configurable resources in the selected CLB may be set or probed.

Since the JBits interface can modify the bitstream it is still possible to create a "bad" configuration bitstream. For this reason it is important to make sure that the circuit which is downloaded to the chip is always a valid design. This is done by only ever modifying a valid design to implement the evolvable hardware. This is a method similar to that demonstrated in Delon Levi's work on GeneticFPGA in [20] where the XC4000 series devices were programmed to implement evolvable hardware. The major obstacle in that implementation was the speed with which the devices could be programmed.

The hardware interface is the XHWIF (Xilinx Hardware InterFace). This is a Java interface which uses native methods to implement the platform dependent parts of the interface. It is also possible to run an XHWIF server, which receives configuration instructions from a remote computer system and configures the local hardware system. This also enables the evolvable hardware to be evaluated in parallel on multiple boards. This interface is also part of Xilinx's Internet Reconfigurable Logic (IRL) methodology, where the computer houses a number of Xilinx parts which can easily be reconfigured through the Java interface. The intention is to create hardware co-processors which can speed up computer systems. Our aim is to use this interface to create evolvable co-processors; evolving solutions to improve the speed of the programs while they are embedded into system designs.

It is possible to partially reconfigure the hardware system using the JBits interface. This is shown diagrammatically in Figure 5.11. This is accomplished by first identifying the particular region of the bitstream which requires reconfiguration and wrapping a header around this bitstream data to identify the address at which to start reconfiguration. During the reconfiguration process the unchanged logic runs as normal with no interruption to the circuit. This means that we can implement the interface to the logic within the circuit, knowing that changes can be made inside the chip without affecting the interface.

In these experiments this interface was used to operate on the bitstream generated by Xilinx Foundation (a circuit design tool from Xilinx). The interface can dynamically modify the circuit design on the Virtex chip which implements the genotype mapping shown in Figure 5.12. The basic element that JBits operates on is the LUT. Each LUT has four input vectors and one output. Since there are $2^4 = 16$ information bits, each controller consists of 22 LUTs*16 bits/LUT = 352 bits. Hence, each individual is represented by 352 bits.

For example, a true table of one four input OR gate is:

OR 1111 1111 1111 1110 0xfffe

The following JBits source code is used to set one LUT to implement an OR gate:

JBits.set(1, 1, LUT.SLICE_G, 0xfffe).

In a Xilinx Virtex V1000 there are 6,127,776 bits required to reprogram the device but we only need to change 16*8 = 128 bits!

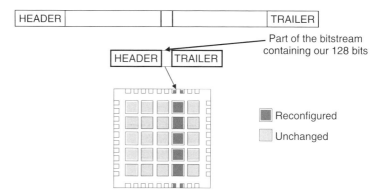

Figure 5.11. Dynamic reconfiguration in action.

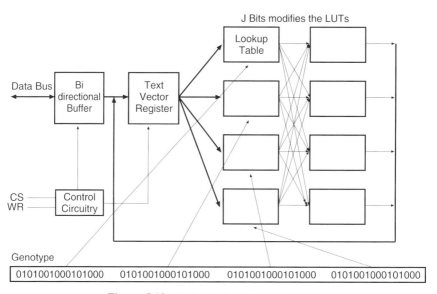

Figure 5.12. A simple baseline circuit [39].

The implementation in this example is based on intrinsic EHW, which is evolved using a Xilinx Virtex FPGA. The genotype of each individual is mapped to the circuit on the chip and the fitness is evaluated on the real robot. The architecture of the control system is depicted in Figure 5.13.

The Khepera robot used in the experiments is shown in Figure 5.14. As shown in Figure 5.15a, the robot has eight infrared proximity sensors and two

EXAMPLES OF EHW-BASED FAULT RECOVERY   **129**

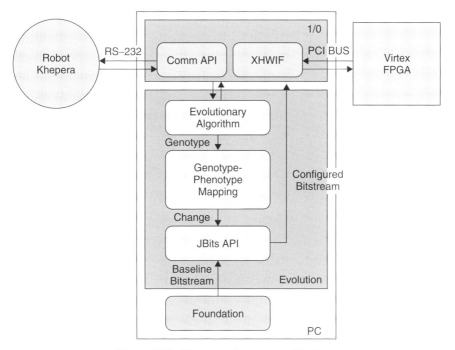

Figure 5.13. The control system architecture.

Figure 5.14. The Khepera robot and its experimental test environment.

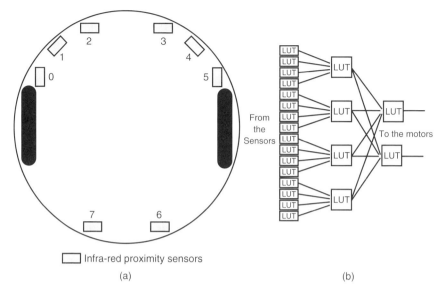

Figure 5.15. (a) The infra-red sensors and (b) the information bits to LUT mapping [25].

wheels [25]. Each sensor can emit infrared light and detect the reflected signals. The sensor value is varied from 0–1023. The higher the sensor values the closer the distance between sensor and obstacle. A value of 400 is used here as a threshold value to convert eight sensor inputs into 8 bits of information. Each wheel of the robot is controlled by an independent DC motor. The controller receives 2 bits of information for four commands: move forward, move backward, turn left and turn right. The robot is controlled by the host PC through an RS-232 link. While this 'link' means that this particular set-up in not completely autonomous (i.e. the PC is still connected), this is simply due to the limitations of the FPGA used in these experiments. Similar results could be obtained by placing a small PIC and FPGA on the Khepera processing stack, or by making use of the new generation of FPGAs, which have on-chip processors included. Such developments would also speed-up the evolution time.

The robot controller is evolved using 22 LUTs (Look Up Tables) on the FPGA, with 8 input bits from the sensors, and 2 output bits to the motors. While this may not be an optimal solution, evolution often finds solutions to problems but not necessarily optimal ones, it will be seen later that the adaptability of such a design has advantages of optimally engineered solutions. Figure 5.15b shows the sensors and the mapping to the LUTs. The hardware interface is the XHWIF (Xilinx Hardware Interface), a Java interface using native methods to operate the hardware platform dependent part of the interface which is used to communicate between the FPGA and the host PC.

The novel EA used in this example is made up of a population of parents and populations of clones (sub-populations) which are used to search for innovation solutions to track dynamic changes in the environment. It is plausible from the engineering point of view that for the innovation population the mutation rate is variable according to the fitness, which reflects the adaptability of the individual to that environment. The mutation process is carried out as follows:

- Individuals with high fitness are subject to a low mutation rate
- Individuals with low fitness are subjected to a high mutation rate.

For the robot navigation problem, a mutation rate defined according to the fitness is given by:

$$n = k(1 - norm\_fit), \quad (5.6)$$

where
  n is the number of bits to be mutated
  k is a constant to be tuned by the user
  $norm\_fit$ is the normalized fitness for the parent population

The evolutionary algorithm manipulates a population of parents and a population of clones. The parent population P is cloned μ times to produce a clone population C. All individuals have the same size as shown in Figure 5.16. The clones were first mutated (which forms a population C') and secondly evaluated (which forms a population C''). The combined population of parents and clones are sorted according to the fitness in descending order. The fittest individuals from P and C'' comprise the population of parents for the next generation while the other individuals are discarded. The evolutionary algorithm used for the Khepera robot experiments is shown in Figure 5.17.

P	C	C	C

Figure 5.16. The structure of the population: parents (P) and population of clones (C).

```
Algorithm:
Initialize a population (P)
evaluate P
while (termination criteria not met)
{
        clone (P) ⇒ μC
        mutate (μC) ⇒ μC'
        evaluate (μC') ⇒ μC''
        selection (P&C'') ⇒ P
}
end
```

Figure 5.17. The evolutionary algorithm used in the Khepera robot experiments.

The fitness measure is a very important part of an evolved robot controller. In on-line evolution, different individuals may not face the same environments. The wheel speeds are determined by the real-time information from the sensors. The controller must send a sequence of different commands to the wheels in real-time. These make the fitness measure much more difficult to evaluate [34]. To solve the problem, the fitness function used in this example did not try to compare the value between input information and output commands. It simply measures the time and distance the robot has run before it hits an obstacle. The longer the time, and the longer the distance the higher the fitness value. Thus the fitness value can be calculated according to a simple relation:

$$fitness = \frac{distance \times time}{1000}. \qquad (5.7)$$

The *distance* is the value of the position counter of the two motors, and the *time* is the value of the I/O period counter. A distance of about 1000 represents a movement of 40 mm. If the robot is turning, there will be no increment in distance value. In order to shorten the time, a time limit value is set to 140. An individual will be killed when the limit is reached; even if it has not hit the wall. At the same time, if the individual is stuck in some position without any distance improvement, it will be killed. The fitness limit in the experiment is 1400, which means the robot moved forward without any steering change. In the experiments, the values of two nearby sensors are used to judge whether the robot is too close to the wall, and an escape time is provided to the individual to quit from the dead zone where the last individual was killed. Since only one robot is used, one individual is running at a time. A sample trajectory of a robot controller evolved in hardware is illustrated in Figure 5.18.

For these experiments, the population of parents had size 16 and was cloned three times. That is, $|P| = 16$ and $\mu = 3$. The population of parents P is not mutated in order to make sure that good individuals in P are not lost by the mutation operation. (This is a type of elitism.) In order to compare the results, about 8 bits in each individual of the cloned populations were mutated—that is, a mutation rate equal to 0.022. Figure 5.19 shows the results for this case. All results presented are for the population of parents and the fitness values are averaged over 10 runs.

The mutation rate is calculated according to Eq. (5.6). Note that, in order to obtain an integer number of bits to be mutated, only the integer part in Eq. (5.6) is taken. Two cases are considered: For $n = 16$, the best individual has normalized fitness 1, therefore no bits are mutated and the worst individual has normalized fitness 0, therefore 16 bits are mutated which corresponds to a mutation rate of 0.045. Figure 5.20 shows the results for this case. For the next case, the mutation rate was increased. For $n = 35$, the worst individual had 35 bits mutated which corresponds a mutation rate of 0.1. The result for this case is shown in Figure 5.21.

In the first case shown in Figure 5.19, considering a constant mutation rate for the cloned populations, it can be noted that it is quite difficult to achieve the

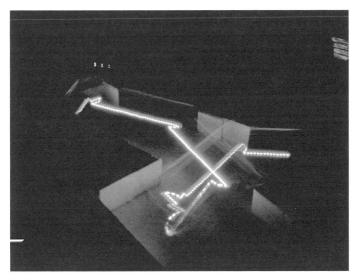

Figure 5.18. Trajectory of robot with an evolved collision avoidance controller.

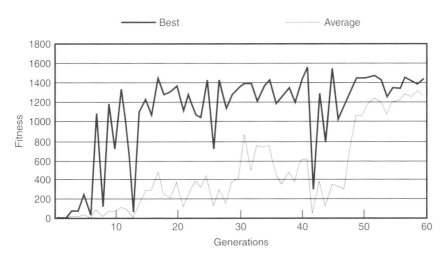

Figure 5.19. Behavior for the best and average fitness of the population of parents (P) when a constant mutation rate was applied to the individuals of the cloned populations (C) [39].

average fitness of 1200 which was reached after 50 generations. In Figure 5.20, where the mutation rate was calculated according to the fitness, the average fitness reached the value 1200 in less than 30 generations with the best individual presenting a quite stable behavior around the value 1400, showing a substantial improvement in performance compared with the former case. In the next case study, shown in Figure 5.21, the mutation rate was increased, that is, for the

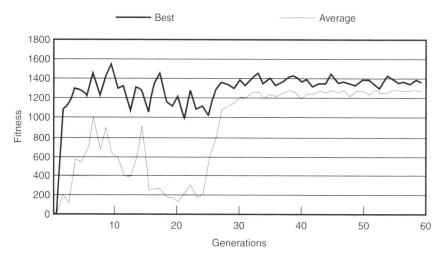

Figure 5.20. Behavior for the best and average fitness of the population of parents (P) when the mutation rate for the individuals of the cloned populations (C) was calculated according to 16*(1-*norm_fit*) [39].

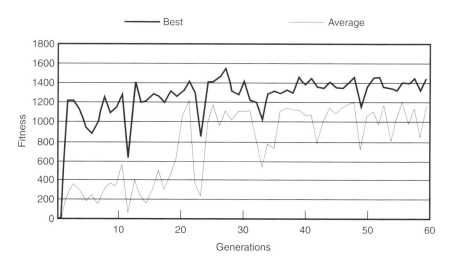

Figure 5.21. Behavior for the best and average fitness of the population of parents (P) when the mutation rate for the individuals of the cloned populations (C) was calculated according to 35*(1-*norm_fit*) [39].

worst individual the number of mutated bits was 35. The behavior of the average fitness is very oscillatory, however, the fitness of the best individual presents a high value around 1400 with more fluctuation when compared with the previous case, which is due to the high mutation rate.

From these experiments it can be observed that the value of the mutation operator is very important to evolve the controller for autonomous robot navigation. It might be argued that it is quite difficult to obtain adaptive behavior with constant mutation rate. With an adaptive mutation rate it was possible to obtain good results with an intermediary mutation rate, which depends on the problem at hand. Also, the experimental results demonstrated that too high mutation rate might lead to an unstable behavior.

The next set of experiments investigated what would happen if sensor faults were injected. For all of these experiments the maximum number of bits of the cloned populations to be mutated is set to $k = 16$. Faults were physically injected by covering one of the robot sensors with a paper mask (see Figure 5.15a). This type of fault fixes the sensor output at 1023. So, could the controller learn how to avoid obstacles with the faulty sensor? The result shown in Figure 5.22 illustrates the evolution when the fault applied to sensor 2 was added before the evolutionary process began. Similar to this case, a fault was introduced in sensor 5 and the result is shown in Figure 5.23. From the results shown, it can be noted that the algorithm is suitable for evolving controllers when a fault is introduced to the sensor of the robot before it is evolved. However, it can be observed that for sensor 2 (frontal sensor) this information has greater significance to the evolution of a successful controller than 5 because when this information is available the fitness of the evolved controller presents a more stable behavior (see Figure 5.20). For sensor 5 (lateral sensor) this information is of less importance to evolve a successful controller because no performance deterioration was observed when this information was available (Figure 5.23).

For the next set of experiments a fault was added when an acceptable controller had already been evolved, that is, during run-time. Faults were added to

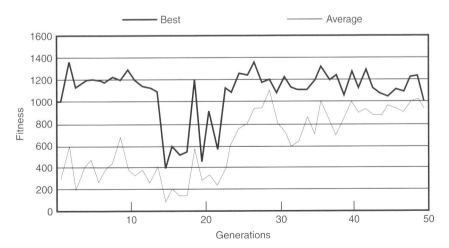

Figure 5.22. Behavior for the best and average fitness of the population of parents (P) when a fault in sensor 2 was added prior to the evolution [39].

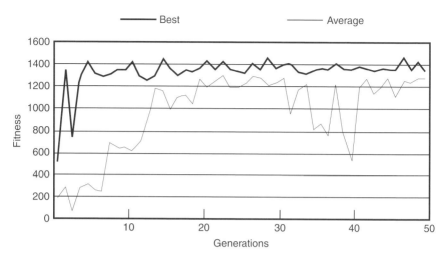

Figure 5.23. Behavior for the best and average fitness of the population of parents (P) when a fault in sensor 5 was added prior to the evolution [39].

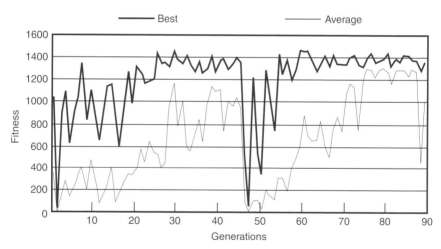

Figure 5.24. Behavior for the best and average fitness of the population of parents (P) when a fault in sensor 2 was added at generation 50 [39].

either sensor 2 or 5 to compare the results. Firstly, a fault was added to sensor 2 at generation 50, results are shown in Figure 5.24. This fault disrupted the controllers of the whole population. It can be observed that both the average fitness and the fitness of the best individual dropped when the fault was added. However, after about 10 generations the fitness of the best individual rises and regains its original value. Similar experiments were carried out when

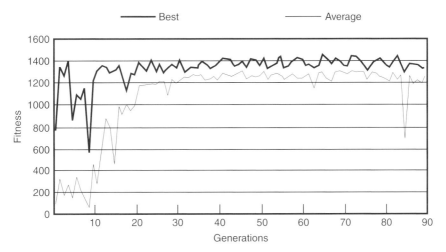

Figure 5.25. Behavior for the best and average fitness of the population of parents (P) when a fault in sensor 5 was added at generation 50 [39].

a fault was added at generation 50 to sensor 5 as depicted in Figure 5.25. The fitness kept its high value without significant change, as if no fault had been added. This suggests that for this particular evolved controller during the first 50 generation the information from sensor 5 is of less importance, i.e. the information provided by the sensor 5 is redundant. Figure 5.24 shows that although this approach does not prevent the failure introduced by the fault (i.e., the drop off of fitness) when it occurs, it does allow it to regain the fitness quite quickly. Another important point is that to evolve a successful controller not necessarily all sensors information (e.g. sensor 5) need be used in this particular case.

This example presented a novel method to evolve FPGA-based controllers for a mobile robot. Experimental results on the Khepera robot have shown that the algorithm is suitable for providing "autonomous" navigation for collision avoidance even with the occurrence of faults in some sensors. A population of parents and cloned individuals, which undergo mutation defined according to the fitness, has been investigated. It was observed that different behaviors emerge, e.g. good adaptation to dynamic changes at intermediary (medium) mutation rates, and the presence of fluctuations, or instability, for high mutation rates. It was also observed that the information of some sensors is redundant and other sensor data are crucial to a successfully evolved controller.

Three important features of an intelligent system have been considered: *autonomy*, *adaptability* and *robustness*:

- *i*) The FPGA allows the reconfiguration of the control circuit, i.e., the controller emerges; it is not pre-designed as is the case in conventional control systems.

ii) By using a mutation rate defined according to the fitness of the individual it is possible to obtain good adaptability in changing environments.

iii) And a robust controller can be evolved, which continues to work even in the presence of faulty sensors.

A more "philosophical" discussion could be started here as to whether intrinsic evolvable hardware is really appropriate for such applications. It is clear that in the early stages of evolution in particular the performance can be quite poor and of course the process kills these poor individuals off (as one might expect from an evolutionary process). While this is acceptable in the laboratory set-up described in these experiments, and in most papers published on this subject, it may not be in a "real" environment. It is clear that further work is required to consider this point and investigate possible solutions to this problem. However, what has been shown by these experiments is that a system can evolve a solution to a particular problem without the need of a human designer. It can, possibly more importantly, continue to evolve throughout the lifetime of a mission and adapt to unexpected events and conditions. This has been demonstrated in this example by sensors failing during operation and observing the recovery to full functionality through the evolutionary process. Such an attribute, while still possessing early poor individuals at least provides some autonomous recovery, adaptation, and would be critical in applications such as unmanned space and underwater missions where the lack of such adaptation mechanisms will simply lead to failure of the mission.

Since the evolutionary algorithm used here is simple, its implementation in an FPGA is straightforward, which means the entire evolutionary process can be done in hardware.

5.7.4 The POEtic Project

The example shown here, and presented to some extent in [4], develops ideas in the framework of a research project, called Reconfigurable POEtic tissue. (The POEtic architecture was described in Chapter 3.) After a short introduction to the POE project for the design of bio-inspired hardware, the example will present an outline of the POEtic device built during the project, focusing on the features used in this example.

Reducing the failure probability and increasing the reliability has been a goal of electronic systems designers ever since the first components were developed. No matter how much care is taken in designing and building an electronic system, sooner or later an individual component will fail. For systems operating in remote environments such as space or deep-sea applications, the effect of a single failure could results in a multi-million dollar installation being rendered useless. With safety critical systems such as aircraft and mobile robotics, the effects are even more severe. Reliability techniques need to be implemented in these applications and many more. The development of fault tolerant techniques was driven by the need for ultra-high availability, reduced maintenance costs, and long life applications to ensure systems can continue to function in spite of faults occurring. The implementation of a fault tolerant mechanism requires four stages [19]:

- Detection of the error;
- Confinement of the error (to prevent propagation through the system);
- Error recovery (to remove the error from the system); and
- Fault treatment and continued system service (to repair and return the system to normal operation).

In this particular example we will concentrate on detection of errors and error recovery. But first we should comment on the new architectural aspects of the POEtic device that make it so amenable to fault tolerant designs. Work has recently begun investigating both evolutionary and developmental approaches to reliable system design in the form of evolvable hardware [38], embryonics [23] and artificial immune systems [5]. The implementation of bio-inspired systems in silicon is quite difficult, due to the sheer number and complexity of the biological mechanisms involved. Conventional approaches exploit a very limited set of biologically-plausible mechanisms to solve a given problem, but often cannot be generalized because of the lack of a methodology in the design of bio-inspired computing machines. This lack is in part due to the heterogeneity of the hardware solutions adopted for bio-inspired systems, which is itself due to the lack of architectures capable of implementing a wide range of bio-inspired mechanisms.

The goal of the POEtic project was the

> "... development of a flexible computational substrate inspired by the evolutionary, developmental and learning phases in biological systems."

Biological inspiration in the design of computing machines finds its source in essentially three biological models [31]: *phylogenesis* (P), the history of the evolution of the species, *ontogenesis* (O), the development of an individual as orchestrated by its genetic code, and *epigenesis* (E), the development of an individual through learning processes (nervous system, immune system) influenced both by the genetic code (the innate) and by the environment (the acquired). These three models share a common basis: the genome.

The POEtic tissue is a multi-cellular, self-contained, flexible, and physical substrate designed to interact with the environment, to develop and dynamically adapt its functionality through a process of evolution, development, and learning to a dynamic and partially unpredictable environment, and to self-repair parts damaged by aging or environmental factors in order to remain viable and perform similar functionalities. Following the three models of bio-inspiration, the POEtic tissue was designed logically as a three layer structure (Figure 5.26 gives an abstract view of this relating to hardware):

- The phylogenetic model acts on the genetic material of a cell. Each cell can contain the entire genome of the tissue. Typically, in the architecture

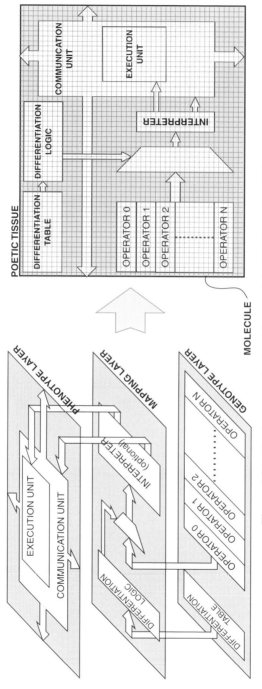

Figure 5.26. The three organizational layers of the POEtic project [40].

defined above, it could be used to find and select the genes of the cells for the *genotype layer*.
- The ontogenetic model concerns the development of the individual. It should act mostly on the *mapping* or *configuration layer* of the cell, implementing cellular differentiation and growth. In addition, ontogenesis will have an impact on the overall architecture of the cells where self-repair (healing) is concerned.
- The epigenetic model modifies the behavior of the organism during its operation, and is therefore best applied to the *phenotype layer*.

Defining separate layers for each model has a number of advantages, as it allows the user to decide whether to implement any or all models for a given problem, and allows the structure of each layer to be adapted to the model. This adaptability is achieved by implementing the cells on a *molecular substrate*, in practice a surface of programmable logic.

Beside this combination of the three main axes of biological self-organization, the tissue presents two more innovative hardware-oriented aspects:

1. A layered hardware structure that matches the three axes of biological organization;
2. An input/output interface with the external world that allows each cell to perceive and modify its environment when and where necessary.

The final hardware design (VLSI device) has a number of specific novel features built into its fabric to assist with bio-inspired designs. It is shown here that these features can also be used effectively for the design of fault tolerant systems. A molecule has eight different operational modes, to speed up some operations, and to use the routing plane (see Chapter 3 for more details on this). Here we describe briefly the different modes. They are described in more detail in [36].

- In **4-LUT** mode, the 16-bit LUT supplies an output, depending on its four inputs.
- In **3-LUT** mode, the LUT is split into two 8-bit LUTs, both supplying a result depending on three inputs. The first result can go through the flip-flop, and is the first output. The second one can be used as a second output, and is directly sent to the south neighbor (or can serve as a carry in parallel operations).
- In **Comm** mode, the LUT is split into one 8-bit LUT, and one 8-bit shift register. This mode could be used to compare a serial input data with data stored in the 8-bit shift register.
- In **Shift Memory** mode, the 16 bits are used as a shift register to store data (for example, as a genome). One input controls the shift while the other input is the data input of the shift memory.

- In **Input** mode, the molecule is a cellular input, connected to the intercellular routing plane. One input is used to enable the communication. When inactive, the molecule can accept a new connection, but won't initiate a connection. When active, a routing process will be launched at least until this input connects to its source. A second input selects the routing mode of the entire POEtic tissue.
- In **Output** mode, the molecule is a cellular output, connected to the intercellular routing plane. One input is used to enable the communication. When inactive, the molecule can accept a new connection, but won't initiate a connection. When active, a routing process will be launched at least until this output connects to one target. Another input supplies the value sent to the routing plane.
- In **Trigger** mode, the 16-bit shift register should contain "000...01" for a 16-bit address system. It is used by the routing plane to synchronize the address decoding during the routing process. One input is a circuit enable that can disable every D flip-flop in the tissue. The other input can reset the routing, which starts a new routing.
- In **Configure** mode, the molecule can partially configure its neighborhood. One input is the configuration control signal, and another one is the configuration shifting to the neighbors.

The mode of a molecule is stored in 3 bits of the configuration.

The configuration system of the molecules can be visualized as a shift register of 76 bits split into 5 blocks: the LUT, the selection of the LUT's input, the switch box, the mode of operation, and an extra block for all other configuration bits. As shown in Figure 5.27, each block contains a specific bit that indicates if the block is to be reconfigured or bypassed. This bit can only be loaded from the microprocessor and it remains stable during the entire lifetime of the organism.

A special configure mode allows a molecule to partially reconfigure its neighborhood. It sends bits coming from another molecule to the configuration of one of its neighbors. By chaining the configurations of neighboring molecules, it is possible to modify multiple molecules at the same time.

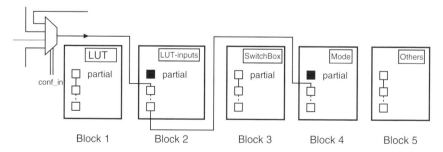

Figure 5.27. Organization of the configuration bits for partial reconfiguration [36].

The POEtic device provides a unique platform for investigating mechanisms at work in biological systems which exhibit fault tolerant behaviors and it is the intent of the following example to demonstrate this through the development of a cellular ontogenetic fault tolerant mechanism on the POEtic tissue based upon growth.

Self-repair, or healing, is a critical mechanism within an organism's response to damage involving the growth of new resources, in the form of cells, and their integration into the organism replacing damaged ones. An electronic system cannot grow new silicon resources in response to device faults in the same way and so growth in silicon is generally emulated by having redundant resources into which the system can grow. The POEtic architecture provides novel features which are particularly useful for implementing models of growth in digital hardware including the underlying molecular architecture, dynamic routing and self-configuration of the tissue.

The example reported here is inspired by two important features of growth, namely *cell division* and *cellular differentiation*. Cell division is a process of self-replication through which cells produce copies of themselves. Cellular differentiation is the process through which cells organize themselves by taking on specific functional types depending upon their positions within an organism. During healing processes some types of differentiated cells in damaged areas within an organism can divide in order to replace those lost, for example mammalian liver cells. This is referred to as renewal by simple duplication. Others rely upon the differentiation of stem cells located within the damaged tissue [1].

Prompted by these distinct modes of growth a novel cell design has been implemented on the POEtic tissue in the context of a test application. Implementation of an embryonic array emulating the processes of cellular differentiation is described here.

In order to investigate issues regarding the implementation of cellular fault tolerant mechanisms on the POEtic tissue a test application previously described in [9]. The test application consists of a cell constructed from nine one-dimensional waveguide mesh elements[8]. This particular application requires real-time audio processing. Cells can be chained together to form a one-dimensional waveguide with length an arbitrary multiple of nine. While this is a specific application executed within the POEtic project, the work reported here is generic and appropriate for any application on one or more devices.

The waveguide cell is shown in Figure 5.28. Figure 5.29 shows a set of input test signals along with the output signals of a correctly operating waveguide cell. These test signals serve as a reference set of waveforms which can be compared against those produced later on when faults are injected.

Biological mechanisms for fault detection in themselves provide a rich field of research to which the POEtic platform is applicable [7]. However as the aim of the cell designs is to investigate growth mechanisms a standard hardware redundancy

---

[8]For those not familiar with waveguide meshes, each mesh element can be considered as a particular type of processing element. The P_s in Figure 5.28 represent values passing between neighboring mesh (processing) elements.

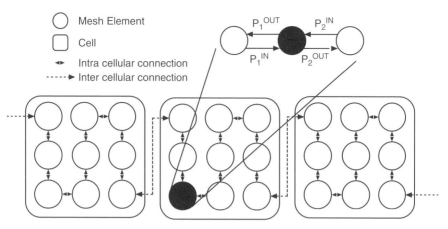

Figure 5.28. Nine mesh-element one-dimensional waveguide cell with left and right input streams [4].

technique has been chosen to provide fault detection in the designs. Figure 5.30 illustrates the hardware redundancy method applied to the test application. It is based upon the assumptions that faults will occur discretely in time and that a fault is only of significance if it causes the cell function at its outputs to deviate from correct behavior. Based upon these assumptions it can be seen that a fault will cause the values at the outputs of cell function copies one and two to differ. This discrepancy is detected by the XOR logic functions comparing the cell function outputs and combined into a fault flag by an OR logic function. The advantages of this method for fault detection are that it is simple, acts at the resolution of a single clock cycle, operates on-line and is applicable to any cell function.

A number of different fault models exist for faults that can occur in programmable VLSI systems such as the POEtic tissue [18]. The important distinction between fault models in this work is that between soft faults and hard faults. Soft faults are disruptions to the data contained within a programmable VLSI device (not its fabric) and can be rectified by resetting or rewriting the device. Hard faults are produced by physical damage to the device and render that part of the device permanently unusable. The hardware redundant fault-detection system described above cannot distinguish between soft and hard faults and so in the following cell designs all faults are assumed to be hard faults. This assumption mandates that the portion of the device which is detected as being faulty be abandoned rather than re-configured. Moreover, it implies that all faults instigate a type of cell apoptosis.

An embryonic array consists of an array of cells implemented in reconfigurable logic each of which contains a set of configuration strings describing every cell function within the system which the cells are to form. This set of configuration strings is analogous to the biological genome contained within every living cell. Each configuration string is analogous to a gene and can be directly translated

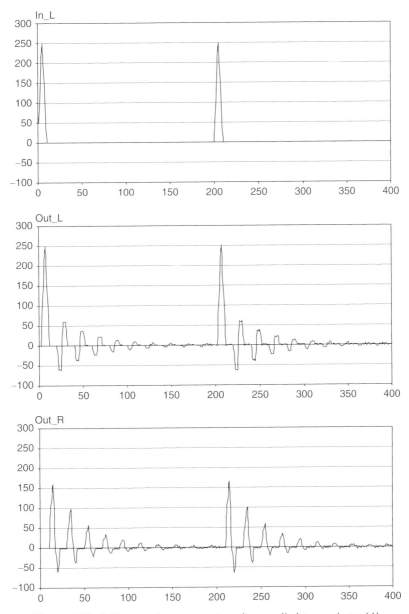

Figure 5.29. Cell output in response to pulses applied to one input [4].

into the cell function it describes in the hardware of the cell. Development of the system is achieved through differentiation during which each cell identifies its configuration string with respect to its location within the array and uses it to configure its function [23, 30], detailed in Figure 5.31.

146 PUTTING EVOLVABLE HARDWARE TO USE

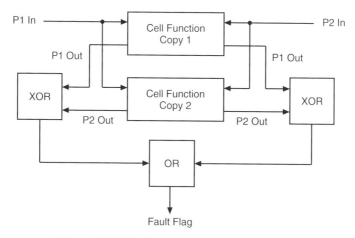

Figure 5.30. Hardware redundant fault-detection.

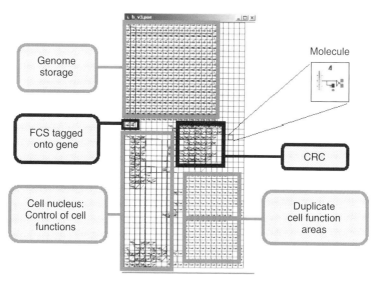

Figure 5.31. Embryonic cell design on the POEtic tissue [4]. (*N.B*: Cell has a single gene in its genome for illustrative purposes.)

Cell function areas are the regions of the cell where the cell application functionality is performed. Duplication of the cell function enables fault detection by the method described above. The areas are initially blank and require configuration from a stored gene. They consist of molecules which have the partial configuration inputs from their neighbors chained together as illustrated in Figure 5.32 and all configuration registers enabled for configuration. This allows an arbitrary cell function to be configured within them. The stored gene therefore consists of the contents of the configuration registers for each molecule in the function listed in the order in which they appear in the chain from the head to the tail.

The stored genome consists of individual gene blocks each of which can be selected by the differentiation system to be the source for configuration of the function areas within the cell. The genes consist of shift memory molecules which store the configuration string in their look up tables (LUTs). The inputs and outputs of the memory molecules are chained together in the same way as the molecules in the function configuration areas. During configuration of the function areas every gene in the stored genome shifts its contents out from its head with the string from the gene selected by the differentiation process being channeled into the cell function areas.

The head of each gene block is looped back into the tail by connecting the memory molecule output of the head molecule to the input of the tail molecule so that the contents of the gene block are retained, illustrated in Figures 5.33 and 5.34.

The use of a stored genome has implications with respect to the redundancy-based fault-detection scheme proposed for the cell design. The stored genome requires a significant area of resources on the tissue which can potentially sustain faults. Approximately four times as many molecules are required to store each gene than are used in the function block that it describes and an embryonic cell will require as many genes as there are different cells in the system. As both function copies are configured from the same stored gene, a fault in the gene will go undetected by the redundancy fault-detection system as both copies will be producing the same erroneous outputs.

A second fault-detection system has therefore been implemented in the embryonic cell design in the form of a cyclic redundancy code (CRC) which can detect faults in the gene being used to configure the cell function by means of a frame check sequence (FCS) which is tagged onto the end of every gene [10]. On configuration of the cell function areas the configuration string for the selected gene including the FCS is passed through the CRC register as the cell function areas are configured. If the stored gene is not corrupted, then the output of all of the CRC register elements will be zero at the end of this process. Otherwise at least one of the outputs of the register elements will be high indicating a fault in the gene.

The cell nucleus is responsible for controlling the five main processes of the embryonic cell. These are cellular differentiation, cell function configuration, fault detection, apoptosis and routing.

1. **Cellular Differentiation**

    Each of the embryonic cells in the array has a differentiation input and output molecule. These inputs and outputs are linked in a chain across the tissue, each cell introducing a one clock-cycle delay between its input and output. On receiving a zero at its input each cell asynchronously sets its output to zero. The first cell in the chain has its input connected to a source external to the tissue into which the differentiation signal is driven. This signal consists of a series of ones equal in length to the number of

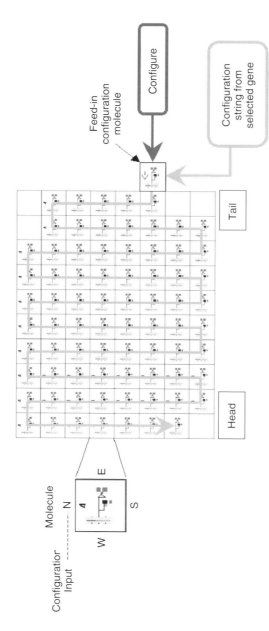

Figure 5.32. Cell function area configuration chain [4].

EXAMPLES OF EHW-BASED FAULT RECOVERY   149

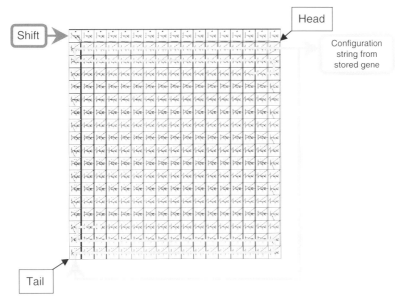

Figure 5.33. Gene block [4].

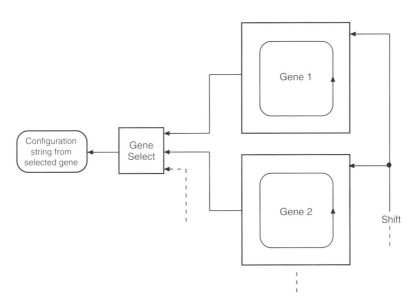

Figure 5.34. Stored genome consisting of selectable gene blocks [4].

cells in the organism being developed. The first cell in the chain therefore receives a series of ones equal in number to this value before receiving a zero. At this point the output of cell one is asynchronously reset to zero causing a chain reaction through which all cell differentiation outputs down the chain asynchronously reset to zero. This terminates the differentiation process in the cells, each of which will have received one less one at its differentiation input than the previous cell in the chain. The cells then select their allocated genes depending upon the number of ones received at their differentiation inputs. Cells which receive no ones at their inputs blank their function areas and are left as unused spare cells. This process can be instigated at any point by simply driving the differentiation signal into the differentiation chain.

2. **Cell Function Configuration**

    Completion of the differentiation process triggers a molecule configuration counter. The counter enables the shift input to each of the gene blocks in the stored genome and enables a configuration molecule which feeds the output of the selected gene into the configuration input of the tail of the function area configuration chain.

    When the counter indicates that the number of molecules in a cell function area have been configured, the enable to the configuration molecule is disabled and the counter resets. At this point the function areas of the cell have been programmed with the selected gene and are ready for integration into the system. Before this can occur however the FCS must be shifted through the CRC register to check that the gene is not corrupted. This is controlled by a second counter which is triggered by the overflow of the first and shifts the gene blocks by a further 32 bits driving the FCS out of the gene block through the CRC register. See Figure 5.35.

3. **Fault Detection**

    The fault-detection techniques implemented in the cell have been described previously. In the cell nucleus the values at the outputs of the two cell function copies are compared and a fault flagged in response to a discrepancy. The integrity of the configuration of these function copies is tested by the CRC register on configuration and if corrupt a fault is flagged. The cell nucleus combines these two fault flags into a single signal which triggers cell apoptosis and differentiation of the system.

    Differentiation in response to a fault is triggered by the faulty cell setting the *mol_enable_out* signal on its trigger molecule low. This is detected by the external system controlling the differentiation signal input to the tissue which drives the signal into the differentiation chain in response.

4. **Apoptosis**

    Apoptosis (i.e., a programmed cell death) in a faulty embryonic cell is achieved by removing this cell from the differentiation process. This is done through a two step process. First, the cell's function must be set to "null", which is accomplished by selecting the blank gene. Second, the cell must be taken out of the differentiation chain, which is accomplished

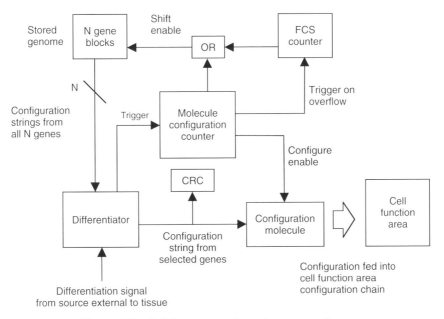

Figure 5.35. Cell function configuration system diagram.

by bypassing the cell. This bypassing eliminates the delay that the cell would otherwise introduce into the differentiation chain. Consequently, the differentiation values received downstream of the faulty cell are shifted by one — that is, a cell that received a chain signal signifying it was "cell n" now receives a signal signifying it is "cell $n-1$". In the cell being removed this action causes the cell to blank its function areas by removing any molecules that may interfere with the operation of the system.

5. **Routing**

Having completed the processes of cellular differentiation and configuration the final step in producing the functioning embryonic system is to route together any input and output molecules which have been configured in the function areas of the cells.

The cell which receives the differentiation value equal to the number of cells in the system, i.e. the first healthy cell in the differentiation chain, is assigned the task of triggering the routing process. Every cell contains a trigger molecule capable of this. On completing the configuration of their function areas every cell sends a pulse on the *start_routing_enable* signal to its inputs entering them into the routing process. This pulse is also sent to the *reset_routing* input on the cell's trigger molecule via a gate. The output of this gate is enabled if it is the first working cell in the chain thereby triggering the routing process. This process ensures that only one trigger molecule will fire at any given time [36].

The cell design described above has been simulated in the presence of randomly generated faults using the POEtic design tool POEticmol [37]. As POEticmol simulates the behavior of the POEtic device using the VHDL description from which the device is fabricated accurate simulations of the effects of faults on system behavior can be made.

After elaboration of the VHDL description of the POEtic tissue POEticmol scans through the entire tissue description compiling a list of signal elements. A number of signal elements are then randomly chosen from this list which will be forced into a fault condition during the simulation. Each signal element is randomly allocated a clock-cycle number from the pre-specified duration of the simulation upon which to become faulty. The fault model used is the "stuck-at" fault, or single hard error (SHE) [18]. The number of faults forced in the simulation is set at a value high enough to guarantee a satisfactory yield of terminal cell faults rather than at a value which is a realistic representation of fault rates for the real device with respect to the test application.

Some example results showing the behavior of the embryonic cell design in response to randomly generated faults can be seen in Figures 5.36 to 5.38. The figures each show the input and output data for two cells, a working cell and a spare cell[9], simulated over five runs of a fixed number of output words. Each run has a different randomly generated set of faults which are to be forced into the tissue. The design is reset after the end of each run and any faults forced into the tissue are removed. Each cell design is simulated under three conditions. In the first simulation no faults are forced into the tissue. This generates the target output which the fault tolerant system is aiming to achieve. In the second simulation faults are forced into the tissue but the fault-detection and growth mechanisms are disabled. In the third simulation the same faults are forced into the tissue with the fault-detection and growth mechanisms enabled.

In Figure 5.37 it can be seen that an unprotected embryonic cell sustains a terminal fault in run 2. In run 2 of Figure 5.38 it can be seen that with the fault tolerant systems enabled the system has detected the induced fault and instigated apoptosis of the faulty cell and re-growth of the system. Data loss in the embryonic system is illustrated by the zero output between points a) and b) produced by the newly grown system. At point a) the system is repaired and fully functional but its response to the input pulse previous to the fault being detected has been wiped by re-growth. By point b) the correct system response to this input has become negligible and a new input pulse stimulates the repaired embryonic system producing incorrupt output data.

An embryonic cellular fault tolerant mechanism has been successfully implemented in simulation on the POEtic tissue. The transparency of the process of mapping an embryonic design onto the POEtic architecture has also been demonstrated. Unlike embryonic implementations on generic FPGA architectures which require complex stages of synthesis and careful tailoring of the embryonic architecture for the target device, compact embryonic designs can be built directly on

---

[9]Both working and spare cells are exposed to faults during the simulations.

EXAMPLES OF EHW-BASED FAULT RECOVERY 153

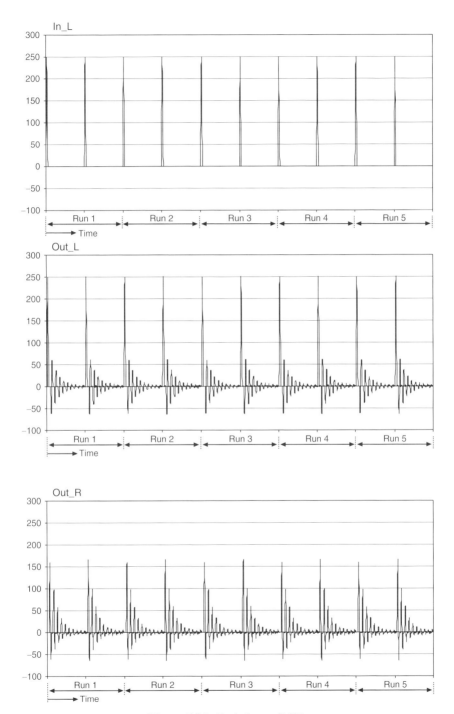

Figure 5.36. Fault-free cell I/O.

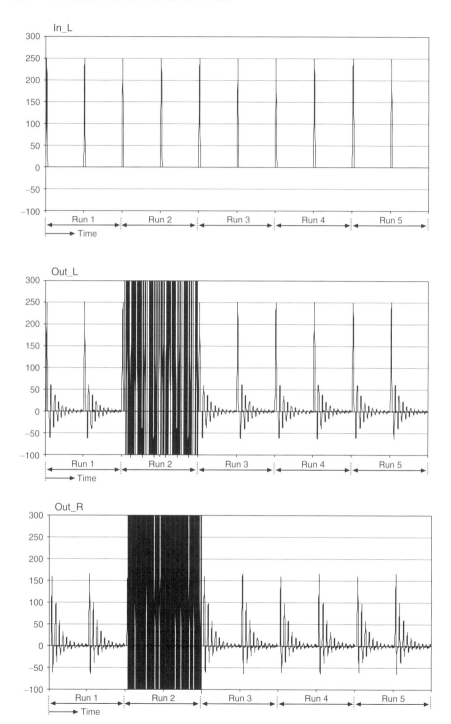

Figure 5.37. I/O with faults. Fault-tolerance OFF.

EXAMPLES OF EHW-BASED FAULT RECOVERY 155

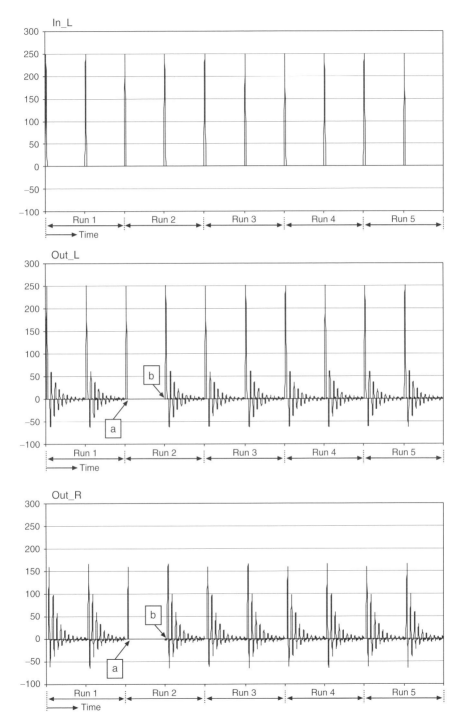

Figure 5.38. I/O with faults. Fault-tolerance ON.

the POEtic tissue at the molecular level. Results of preliminary simulations in the presence of randomly introduced faults show that the cell design is capable of successfully detecting, repairing and recovering from terminal faults in cell function.

The POEtic device, as illustrated with a specific example here, was the development of a hardware platform capable of implementing bio-inspired systems in digital hardware. In particular, the final hardware device, while similar to conventional FPGAs in that it is a reconfigurable logic device capable of implementing any digital circuit, was designed with a number of novel features which facilitate the realization of bio-inspired systems in general and of fault tolerant circuits in particular. These features include dynamic reconfiguration and on-chip re-programming. This example has considers a number of the novel features designed into the ASIC in the context of fault tolerant system design and has shown how an ensemble of different, but often complementary, techniques have been implemented using this novel device. While this example has shown the effectiveness of the POEtic device in the specific area of fault tolerance, the possibilities of the device for research in bio-inspired hardware is not limited to this area: the POEtic device, for the first time, gives researchers in the field of bio-inspired engineering a real chance to evolve, adapt, develop, grow hardware systems. Researchers no longer have to rely strictly on simulations.

5.7.5 Embryo Development

This final example, while still considering evolvable hardware, considers another biological process, that of embryo development. Evolution and development are combined in this example. We will see how evolution and development complement each other and might in the future help with the scalability problems that evolvable hardware sometimes suffers from. However, before talking about the actual example and the processes involved, let's spend a little time considering the biology.

Multi-cellular organisms are the most advanced type of creatures which have evolved over billions of years of evolution. They possess several intrinsic characteristics that electronic engineers earnestly long for, in particular, growth and fault-tolerance. These features are achieved by means of differentiating what are initially identical cells, all of which are developed from one special cell (zygote). The entire process, called development, is controlled by the interaction of cells rather than by a centralized process. (Such decentralized systems are also of interest to engineers).

The development of an embryo is determined by its genes, which control where, when and how proteins are synthesized [41]. Complex interactions between various proteins and between proteins and genes within cells and hence interactions between cells are set up by the activities of the genes. These interactions control how an embryo develops.

Biological development involves several key aspects: cell division, emergence of pattern, change in form, migration, cell differentiation and growth. Cell differentiation emerges as a result of differences in gene activities which lead to the synthesis of different proteins. As development is progressive, the fate of cells becomes determined at various times within the development process of an individual. Inductive interactions by means of chemicals or proteins between cells can make cells different from each other (i.e., differentiate) and the response to these inductive signals depends on the states of cells. Patterning can involve the interpretation of positional information and lateral inhibition.

Development has inspired several hardware research projects in the past [6, 12, 24]. However, in this example a new development-inspired technique [22] will be considered that makes use of "chemical signals" which, among other things, grants the system high tolerance to transient faults.

One of the most fundamental features of development is the universal cell structure. Each of the cells in a multi-cellular organism contains the entire genetic material of the organism, which is the genome.

In the digital hardware model proposed, shown in Figure 5.39, every cell has direct access only to the information of its four adjacent cells. That is, no direct long distance interaction between non-adjoining cells is permitted.

The internal structure of digital cells is shown in Figure 5.40. In this model a digital cell is composed of three main components: Control Unit (CU), Execution Unit (EU) and Chemical Diffusion module (CD).

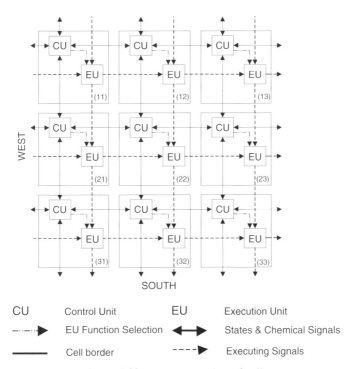

Figure 5.39. Inter-connection of cells.

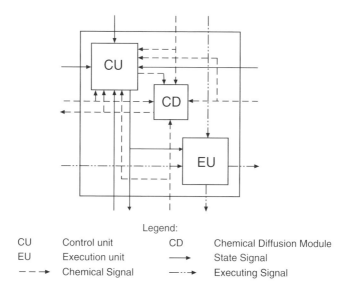

Figure 5.40. Digital cell structure.

The Control Unit (CU) has a States Register, which stores the internal states of the cell, including the cell state (type) and chemical values. Each CU connects to its 4 immediate neighbors (shown in Figure 5.39) and a Next States and Chemical Generator (NSCG) determines its own next state/chemicals according to the current states and chemicals of the neighbors, its own state and its own chemical values (illustrated in Figure 5.40). The NSCG contains two components: Next States Generator (NSG) and Next Chemical Generator (NCG), both of which, in this particular case, are built from combinational circuits.

The EU Function Selection signal (the state of a cell) is 2-bits wide: 0 denotes a dead cell, in this case the EU will simply propagate its west (left) inputs to its south and east neighbors. All other cell types denote that this is a living cell for which the EU will execute and propagate its calculated output to the south and east.

The Execution Unit (EU) is the circuit incorporated to do the real calculation of the target application. The inputs to each EU come from its immediate west and north neighbors, and the state of this cell (refer to Figure 5.40). Every EU also propagates its output (Executing Signals) to its immediate south and east neighbors. The Execution Unit Core (EUC) is the evolvable core logic circuit, which determines how to process the input signals in the EU.

In this example only combinational applications are considered—hence the EU is a combinational circuit. However, this could easily be extended to sequential applications. The state and chemical signals are 2-bits and 4-bits wide respectively, while the width of the Executing Signal is 3-bits. Both the internal core logical structures of the EU (EUC) and the CU (NSG and NCG) are determined through evolution on the hardware. The genotype encodes both the EU and CU internal structures using a cartesian genetic programming compatible representation [27]. More specifically, the genotype is a program expressed as an indexed graph encoded as a linear string of integers. This means the genotype contains a list of node connections and functions.

The Chemical Diffusion module (CD) mimics aspects of the real environment in which biological organisms live. In principle, the CD should not be a component of a digital cell. However, hardware design decisions make it more convenient practically, so it is merged into the cell's internal structure in this implementation.

The chemical signal is introduced to transmit information between cells. The chemical signal also serves as a source of "energy" to transform a dead cell into a living cell.

Previous experiments [26, 28] suggest that chemicals are indispensable in order to achieve robust solutions. Without chemicals, evolved individuals have poor stability and much lower fitness. Chemical diffusion regulation is the key mechanism in this process. Cells have a means to send long-distance messages. The chemical diffusion rule employed in this work is similar to that in [26], except that there are only four immediate neighbors in this case. The rule thus becomes:

$$(C_{ij})_{t+1} = \frac{1}{2}(C_{ij})_t + \frac{1}{8}\sum_{k,l \in N}(C_{kl})_t. \tag{5.8}$$

Let N denote all the four immediate neighbors of a cell at (i, j) with neighboring position (k, l). The chemical at this position, at the new time step, is given by Eq. (5.8). This equation ensures that each cell retains half of its previous chemical and distributes the other half equally to its four adjacent cells and receives the diffused chemical from them. It is obvious from this rule that chemicals are conserved—apart from the unavoidable loss when the level falls below one—in the diffusion procedure.

The main task of the Chemical Diffusion module (CD) (in Figure 5.40) is to calculate the diffused chemical based on the chemicals from the four immediate neighbors and the cell's own chemical value. The CD also propagates the calculated value to the four adjacent cells.

Given a genotype, the inner-structure of the cells is determined and a zygote (initial live cell) is placed at the center of an "artificial environment" with x rows and y columns of cells. Initially apart from the zygote cell all cells are dead (in state 0). The position of the zygote was selected to speed up the growth process: it takes least time for the digital organism to "cover" the entire area if the zygote is arranged in the center. The inputs to the cells on the border of this environment are fixed to 0. In the model cells require the presence of chemicals to live. This means that initially some chemicals must be "injected" into the zygote cell.

Given a genotype, the growth procedure is described as follows:

1. Initialize the chemical and the zygote
2. Chemical diffusion

3. All state cells update their states simultaneously
4. If there is no chemical at a position or all the cell's four neighbors and itself are dead, then this cell's internal program will not be executed
5. Otherwise, it executes the program that is encoded by the genotype to generate its next time chemical and state based on current states and chemicals
6. If the next state generated is alive, then overwrite the chemical at this position with its own generated chemical. Otherwise, do not touch the chemical at this position
7. If the stopping criterion is not satisfied, go to 2

This model was inspired by the software simulation of the "French Flag Problem" described in [28]. In that work, a 63-cell (9-by-7) sized French flag was the intended final pattern. Two bits were used to represent the states of cells. 0 represented a (dead) cell without any colors (gray in pictures); 1, 2, 3 were blue, white and red cells, respectively, which were all stem cells. One 8-bit wide chemical was used. There was no EU since the organism function involved coloring[10]. Once successful organisms were evolved, it was demonstrated that some of these organisms could recover automatically from multiple induced faults.

In this example an application was chosen to give an additional functional output, which in this case this was a 2-bit multiplier. The task was to evolve a cellular circuit that would grow to become a 3×3 cell organism that implements a 2-bit multiplier. The inputs to this multiplier were connected to execution signals of cells (1, 1) and (2, 1), while the output execution signals of cells (2, 3) and (3, 3) drove the output result of this multiplier.

We will now describe in some detail the hardware implementation of these elements. The core of the Intrinsic Evolvable HardWare (IEHW) system is the fitness evaluation module. Digital cells, each of which contains one or more evolvable sub-logic circuits, are the building blocks of the digital organism. The molecules are the lowest level elements of the evolvable components of this model. Indeed, they are the fundamental gates that make up the genome of the cell. Each evolvable sub-circuit is composed of several molecules.

In order to save resources, the genotype is stored centrally in registers outside of the digital organism. Each molecule inputs, including the input selection signals and function selection signal ($I1$, $I2$, $I3$ and *Func* in Figure 5.41), are connected to the corresponding bits in the genotype. The *Data* input pin connects to all the input data available to this molecule, which may include inputs to the entire circuit or the outputs of the molecules on the left of it.

Due to limited resources in the intended hardware platform, the width of the data signal is set to 8. This means each input selection signal is three bits wide. This raises a question about what basic set of digital hardware components are available to the evolutionary process. It was decided that a 3-input

---

[10] In the coloring problem all you need do is represent blue, red and white—i.e., only provide the state information. With a 2-bit adder you have state, obviously, but you also need functional information to do the 2-bit add. Hence, in the 2-bit adder an EU is needed, whereas in the coloring problem an EU is not needed.

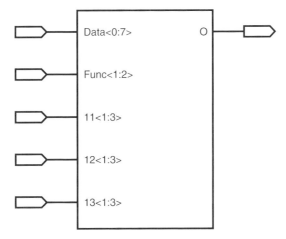

Figure 5.41. Molecule interface.

Universal-logic-module would be suitable. In this particular case we used a 2-1 multiplexer[11]. Although other higher order ULMs could have been used, it was found during the experiments that this particular MUX fits the average fan-in requirement of human-designed circuits and therefore was deemed appropriate for these experiments. If larger fan-in cells were employed the wiring density/complexity and wiring channel-to-processing proportion will increase dramatically [8]. The MUX specified in [8] is identified as an appropriate candidate for general-purpose fundamental logic elements. The MUX defines a three variable output function f

$$f = y_1 y_2 + \overline{y}_1 y_3.$$

This configuration, along with negation of the input variables and the constants 1 and 0, can realize all 2-input Boolean functions. Consequently, the functions available to evolution in a molecule are limited to four types of multiplexers (as in [26]), which are shown in Table 5.5. Since we have only four functions to choose from, two bits are sufficient for the signal *func*.

The available inputs to a molecule in the hardware implementation will be constrained. As a result, no fixed logic 1 or logic 0 are available as inputs to the molecules. However, 1 and 0 can be achieved directly and efficiently by using a MUX. That is, a fixed logic 1 can be achieved by using MUX3(X, X, X) and a fixed logic 0 can be achieved by MUX2(X, X, X) where X refers to one input variable to a molecule (e.g., A or B or C from Table 5.5.)

$I1$, $I2$ and $I3$ (in Figure 5.41) control three 8-to-1 multiplexers to select the inputs from the 8-bit width input Data. The selected inputs are then fed to each

---

[11] an n-input universal logic module (ULM) is logic function which is capable of implementing any function with n-1 independent binary input variables [8].

TABLE 5.5. Available Functions for Molecules

Name	Algebraic Expression
MUX1(A,B,C)	$A \cdot C + B \cdot \overline{C}$
MUX2(A,B,C)	$\overline{A} \cdot C + B \cdot \overline{C}$
MUX3(A,B,C)	$A \cdot C + \overline{B} \cdot \overline{C}$
MUX4(A,B,C)	$\overline{A} \cdot C + \overline{B} \cdot \overline{C}$

of the four MUX types (MUX1 to MUX4). One of the four outputs of these MUXs will be selected by the *Func* input as the output O of this molecule.

This infrastructure of molecule has considerable similarity with the multiplexer-based architecture of a Xilinx FPGA, but with less flexibility. This is a trade-off to eliminate the possibility of evolving any configuration that could lead to permanent hardware damage.

The external interface of the digital organism is shown in Figure 5.42. Pins "A", "B" are the inputs and pin "Result" is the output of the 2-bit multiplier. Clk is the global clock signal. If the Reset pin is high, all the internal registers will be set to their initial values. All of the remaining pins are dedicated to allow the injection of transient fault(s) into the digital organism. When the InjectFault pin is high, Value will be written into the chemical of cell at coordinate (X, Y) if VTYPE is low. Otherwise the lowest two bits of Value are written into the state of the cell. During this process the whole organism stops any growth process.

Every cell has an identical structure. Figure 5.43 shows the external interface of a digital cell. Pins InjectFault, VTYPE and Value are connected to their global counterparts. If this cell is at the coordinate (X, Y) and InjectFault is active (high), the CS pin of this cell will be driven to high and the cell will overwrite

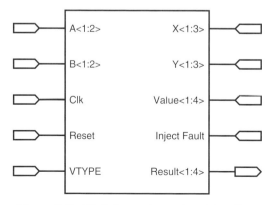

Figure 5.42. Digital organism external interface.

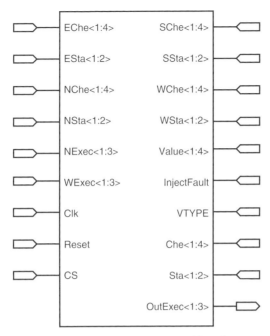

Figure 5.43. Digital cell external interface.

its own chemical or state with Value. A growth step lasts two clock cycles. At the falling edge of the first clock cycle a live cell (its state is not 0) will replace the existing chemical with its generated one. At the rising edge of the second clock, the chemicals diffuse according to the rule described earlier (Eq. (5.8)). During the rising edge of the first clock cycle in the next growth step the state will be updated.

In the work presented here we used a software simulation to evolve the desired structures. In order to simplify the implementation, two phases are designed to evolve the entire digital organism:

1. Evolve the structure of the EUC and the distribution of states of the 3×3 cells. The genotype in this phase consists of two parts: a CGP part which encodes the EUC structure and a second part to encode the states for all nine cells. The fitness is the number of correct bits of the multiplier result output. It has two 2-bit inputs, so there are $2^2 \times 2^2 = 16$ possible combination of inputs. Since the output of a 2-bit multiplier is 4-bits wide, $16 \times 4 = 64$ is the maximum fitness for the organism in a given step.
2. The structures of the NCG and the NSG will be evolved to discover a stable solution for the states distribution of cells found in the first phase. This phase is the same as the evolution process described in the French Flag problem (apart from some parameter values) [26].

TABLE 5.6. Chosen Cell State Pattern

0	1	3
2	3	2
3	2	3

One of the patterns found in the first phase is shown in Table 5.6. This pattern utilizes most available cells with a diverse and complete distribution of states, so it was chosen as the target configuration of the digital organism along with its corresponding EUC structure obtained via evolution.

In order to search for more resource-efficient individuals, once a perfect solution was found, the evolutionary process would receive a fitness bonus based on how many molecules it used, where the fewer the better.

The structures of the evolvable sub-circuits were evolved in software and a robust solution was applied to the hardware. The FPGA implementation was synthesized using ISE 6.1i from Xilinx and downloaded into the hardware. A set of detailed waveforms is illustrated in Figure 5.44. It can be seen that the organism matures at 1 ns, when the state pattern is identical to that shown in Table 5.6.

In the next experiment enough time was allocated to allow the organism to grow and mature. Subsequently, two sets of transient faults were injected. The first set was composed of four transient errors in the chemicals of cells (2, 1), (2, 2), (2, 3) and (3, 3). The second set of faults were injected into the states of cells (2, 1), (2, 3) and (3, 1). Every fault was chosen to make the corresponding value (chemical or state) 0. The time between the injections of the two sets of transient faults was long enough for the organism to recover completely and to stabilize itself both in terms of chemical and state values.

The recovery from of the first set of transient chemical faults is illustrated in Figure 5.45. At the beginning of this experiment the chemical values of some of the cells were modified and then the organism resumed its growth cycle. It recovers at 2.4 ns and the output returns to the correct value.

Figure 5.46 demonstrates the recovery procedure from the second set of experiments, which this time injects transient faults to the cell's state value. The states of the three selected "victim" cells were killed (state 0) at the beginning of this period. The organism again recovers to the correct pattern at 4 ns.

The FPGA used in these experiments, hosted on the RC1000 board [42], connects to the host PC using a data width of 8-bit read and 8-bit write. A further FPGA module "IOControler" was implemented to latch all of the required inputs and feed them to the digital organism. Another function of the IOControler is to cache the result output of the digital organism.

After implementing the digital organism, several other functional modules which are indispensable in the Intrinsic Evolvable Hardware platform (IEHW) were identified. The top level modules are depicted in Figure 5.47.

The IEHW implementation includes three main outputs: MAX_Fitness, GenerationCount and Genotype. The first and the second of these will be updated

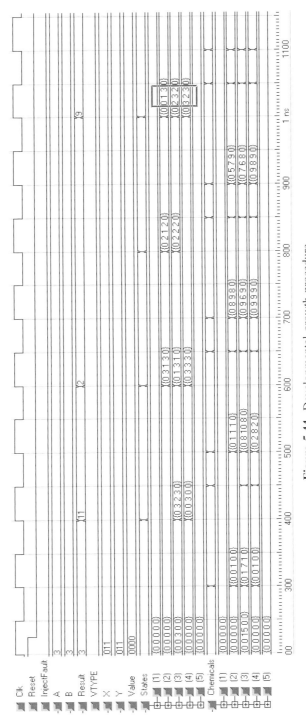

Figure 5.44. Developmental growth procedure.

165

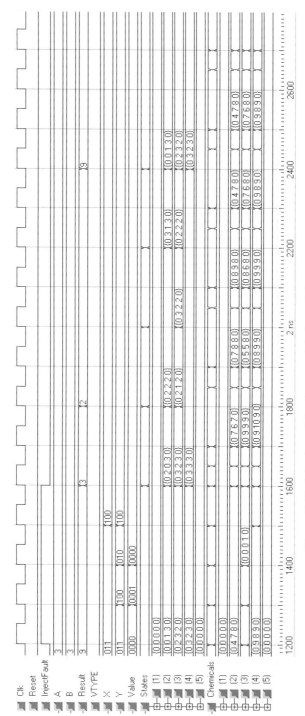

Figure 5.45. Injection of the first set of faults and the recovery procedure.

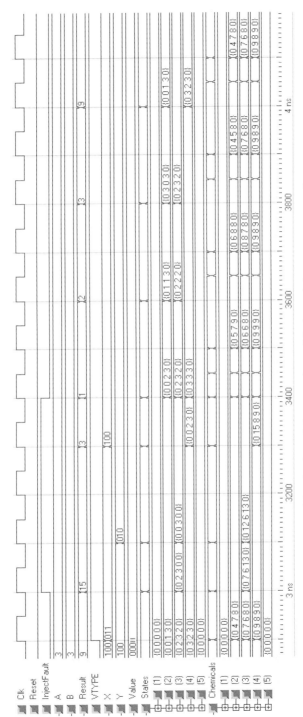

Figure 5.46. Injection of the second set of faults and the recovery procedure.

167

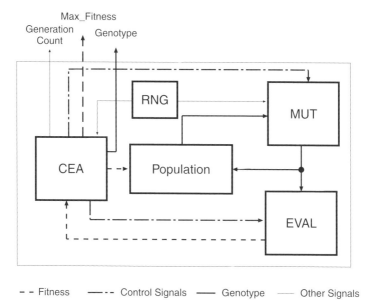

Figure 5.47. Top-level overview of the intrinsic evolvable hardware platform [22].

every generation to reflect the latest values, while the last output always propagates the best individual that is evolved so far. Only two inputs are required for this IEHW to function: the global clock signal and reset signal. Other inputs are optional parameters, such as the seed for RNG and a stop fitness signal.

Before describing the modules in detail, the evolutionary algorithm employed in this example will be discussed first. Most existing widely employed evolutionary algorithms are designed primarily for software implementation, so they are not particularly hardware friendly and efficient—or in some cases even doable.

One of the significant hindrances of transforming the most popular EAs to hardware is the sorting issue. Most popular EAs require some kind of sorting to work properly. Taking the $(\mu + \lambda)$-evolution strategy [32] as an example[12]. It selects the best individuals from among the μ parents and λ offspring to generate the population for the next generation. This mechanism infers that the fitness for all offspring and parents has to be sorted to determine which are the fittest individuals. However, hardware implementation of sorting is not only inconvenient, but also extremely expensive in terms of both resources and design complexity.

In order to avoid this and other issues of the more popular EAs and increase the efficiency of evolution, other alternative evolution strategies were investigated. D. Levi's HereBoy [21] algorithm was considered a good candidate since it was designed specifically for a hardware implementation (on FPGAs).

HereBoy is a combination of features taken from GAs and simulated annealing (SA). The binary chromosome (a string of 1s and 0s) is the data structure

---

[12]Only those ES in which μ is greater than 1 are considered here.

in HereBoy. In principle this kind of chromosome can be mapped to any problem domain. The population only contains one individual and mutation is the only variation operator employed. As done in SA, during each iteration the chromosome is mutated by flipping bits and then evaluated. A better fit offspring always replaces a worse fit parent, but a worse fit offspring may still be kept with a low probability p_r. This means that sometimes a worse chromosome will be maintained. This mechanism allows the system to search for better solutions because it provides an escape from local optima in the fitness landscape.

It is obvious that no sorting is required to perform a HereBoy type algorithm. Inspired by this algorithm, a new EA as follows was conceived:

1. Randomly initiate n individuals.
2. Evaluate all of them.
3. Mutate every individual once to generate n offspring (Mutation rate p_m is fitness related, described below).
4. Evaluate all offspring.
5. If an offspring (F_o) is better than its parent (F_p), replace its parent with it.
6. Else if an offspring is worse than its parent, the offspring has a probability p_r to substitute its parent.
7. Else (when an offspring's fitness is the same as its parent), the offspring has a constant probability p_e (which is determined as an input parameter beforehand) to replace the parent.
8. Unless the stopping criterion is reached, go to step 3.

The flow chart of this algorithm is illustrated in Figure 5.48. The main difference from the original HereBoy is that a population with more than one individual is possible in the proposed algorithm: HereBoy is a special case of this algorithm which has only one individual in its population.

An adaptive mutation rate has been shown to be efficient for hardware evolution. For instance, in [17] a mobile robot could adapt to unpredictable environments with the help of an evolutionary algorithm that employed a mutation rate defined according to the normalized fitness. It was also suggested in [21] that the evolution benefits from an adaptive mutation rate. Based on these findings, an adaptive mutation rate p_m was employed in this work. It is defined as:

$$p_m = \begin{cases} p_{\min} & \text{if} \quad p_c < p_{\min} \\ p_c & \text{if} \quad p_{\min} < p_c < p_{\max} \\ p_{\max} & \text{otherwise} \end{cases} \quad (5.9)$$

p_c is calculated based on the individual's current fitness f and the maximum fitness $f_{\max}$, given as:

$$p_c = p'_m \left(1 - \frac{f}{f_{\max}}\right), \quad (5.10)$$

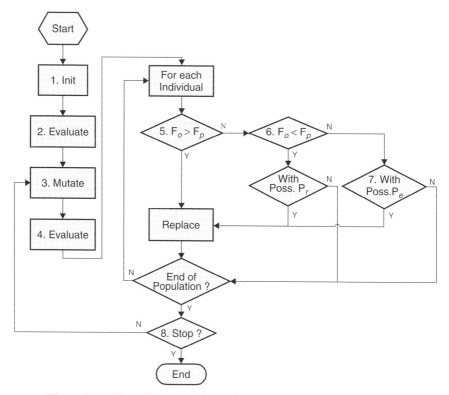

Figure 5.48. Flow diagram of the evolutionary algorithm proposed [22].

where p_{min}, p_{max} and p'_m are predefined parameters. In general, p_{min} multiplied by the number of total molecules should be greater than or equal to 1 and p_{max} should be equal to or less than 0.5.

p_m starts at a high value (normally p_{max}) when the evolution begins but declines to p_{min} as the process converges on the final solution. This scheme allows the algorithm to focus on searching for generally good solutions at the beginning and then fine tunes them to evolve the best one.

The probability p_r of a worse offspring replacing its parent is governed by a similar rule:

$$p_c = p'_r \left(1 - \frac{f}{f_{max}}\right) \qquad (5.11)$$

p'_r is another input parameter, called the *principle maximum mutation rate*. The other part of the product which generates the p_r is a fraction which will reduce from 1 to 0 as the run converges. So p_r decreases as the fitness of individual approaches the maximum. p_r is introduced to help escape from a local optimum.

Referring to Figure 5.47, there are five functionally independent top-level modules which implement the IEHW system. All the genotypes of each individual are stored in the *Population module*. This is implemented in the FPGA as distributed RAM because only one individual is manipulated at any given time.

The Controller of EA (CEA) is the central management component which supervises the entire evolution process and all the other modules. The fitness for all the individuals in the population is also stored in this module. The EA is essentially the same as described earlier, with one exception: no adaptive mutation rate is employed, although an adaptive replacement rate of a bad-offspring overwriting its better parent is used. The CEA module is realized as a finite state machine (FSM). It is obvious from Figure 5.47 that the CEA has nothing to do with any genotype signals, so it is a representation-independent functional module.

RNG is a 64 bit Linear Feed-back Shift Register (LFSR) (more details are available in the references [32] and [11]), which is employed as a pseudo-Random Number Generator [2]. If supplied, the seed of RNG will also be saved in this module.

The main function of the Mutation Module (MUT) is to mutate a given genotype and latch the mutated genotype to be used by the EVAL. This module reads in the mutation rate and mutates molecules one by one until the specified mutation number is met. This module is also implemented as an FSM.

The core component of this IEHW is the Evaluation Module (EVAL), where the digital organism resides. Its main function is to evaluate the fitness of each individual. This module feeds every possible input to the 2-bit multiplier implemented by the evolved digital organism and sums up the total number of correct bits. Finally, the result of the difference between the total correct bits from the individual and the maximum possible correct bits (64 in this case) is the fitness of this individual. The EVAL module is made up from two FSMs. One FSM is used to manage the digital organism while the other FSM is in charge of feeding the inputs and calculating the correct output bits. It also performs the final subtraction to generate the fitness output of the module.

Rather than a central bus, all the modules have their dedicated input/output signals as shown in Figure 5.47. This feature improves the efficiency of the evolution process.

After the reset signal is pulsed (low) for one clock cycle, all of the modules and all FSM and internal registers are forced to their initial states. In this state, the IEHW will receive and latch any input parameters if provided, otherwise the default parameters are used. When the start signal is activated by the host PC the CEA module takes over for the IEHW.

Firstly, the population are initiated one by one, evaluated and saved into the Population Module. The CEA signals the MUT to mutate at the highest possible rate so all the molecules in the genotype are randomly generated, then EVAL evaluates it and propagates the fitness to CEA. Finally the CEA saves the fitness and signals from the Population Module to store the newly generated individual. These individuals make up the 0 (first) generation.

Secondly, after the initial population is ready, the main loop of the evolution process begins. In each generation, the CEA selects the individuals from the Population Module and feeds them to the MUT. The mutated genotype is then evaluated by the EVAL module, and the fitness is again propagated back to CEA. If the offspring is to replace its parent, then the CEA asks the Population Module to store the mutated genotype. Otherwise the content of the Population module is unchanged. After all the individuals have undergone this procedure, a new generation is created. The evolution continues until a specified number of generations are processed or an individual with a high enough fitness was found.

When the main loop of the evolution process terminates, the best individual evolved is presented through the Genotype pin, while its fitness and the generation where this evolution stops are propagated out via the MAX_Fitness and GenerationCount signals respectively.

The biological development model proposed in this example is capable of exhibiting the intrinsic highly fault tolerant feature similar to its living organism counterpart when it is applied to real world application. In fact, typically the best solutions discovered are able to tolerate virtually any transient damages.

The 2-bit multiplier example is a simple problem, but it is sufficient to serve as a proof of concept. Nevertheless, when this model is applied to more complex applications, its development will most likely be more efficient than conventional approaches. Just as is the case in nature, human beings are too complicated to be described in detail, while DNA wisely encodes the development procedure of a person. In the future this model can be applied to more sophisticated systems without fundamental modification. With the help of the IEHW platform described, the evolution process should considerably faster than a VHDL or Verilog development.

5.8 REMARKS

In this chapter we presented a number of examples, from both analog and digital problem domains, to demonstrate the power of EHW. The applications ranged from simple digital logic design, to continuous control problems, to signal processing problems. We hope that these examples have given you practical and thorough understanding for what EHW can do and some of the issues to consider when applying EHW techniques to your own real-world problems.

One of the more important points is COTS devices suitable for EHW applications are readily available. The field of possible applications is only expected to grow as more sophisticated reprogrammable devices—particularly in the analog realm—begin to hit the marketplace.

In the final chapter, we will summaries some of the issues facing the world of evolvable hardware and give some opinions on where we think are the hot, and not so hot, areas for future research efforts.

REFERENCES

1. Alberts B, Bray D, Lewis J, Raff M, Roberts K, and Watson J 2002, *Molecular Biology of the Cell*, Garland Science
2. Alfke P 1996, "Efficient Shift Registers, LFSR Counters, and Long Pseudo-Random Sequence Generators", *Xilinx Application note* XAPP 052 (Version 1.1)

3. Ball H and Hardy F 1969, "Effects and detection of intermittent failures in digital systems", *AFIPS Conference Proceedings*, Vol 35, 329–335
4. Barker W, and Tyrrell A M 2005, "Hardware Fault Tolerance within the POEtic System", *6th International Conference on Evolvable Systems*, LNCS 3637, 25–36
5. Bradley D W and Tyrrell A M 2002, "Immunotronics: Novel Finite State Machine Architectures with Built in Self Test using Self-Nonself Differentiation", *IEEE Transactions on Evolutionary Computation* 6(3), 227–238
6. Canham R and Tyrrell A M 2003, "An Embryonic Array with Improved Efficiency and Fault Tolerance", *5th NASA Conference on Evolvable Hardware*, 265–272
7. Canham R, Tyrrell A M 2002, "A Multilayered Immune System for Hardware Fault Tolerance within an Embryonic Array", *First International Conference on Artificial Immune Systems*, 3–11
8. Chen X and Hurst S L 1982, "A comparison of universla-logic-module realizations and their application in the synthesis of combinatorial and sequential logic networks", *IEEE Transactions on Computers*, Vol. C-31, 140–147
9. Cooper C, Howard D M and Tyrrell A M 2004, "Using GAs to Create a Waveguide Model of the Oral Vocal Tract", *6th European Workshop on Evolutionary Computation in Image Analysis and Signal Processing*, 2880–288
10. Costello D and Lin S 2004, *Error Control Coding*, 2nd Ed. Prentice Hall, 136–188
11. Golomb S 1967, *Shift Register Sequences*, Holden-Day, San Francisco
12. Gordon T, Bentley P 2002, "Towards development in evolvable hardware", *Proceedings of the NASA/DoD Conference on Evolvable Hardware*, 241–250
13. Haddow P. and Tufte G 1999. "Evolving a Robot Controller in Hardware", *Proceedings of the Norwegian Computer Science Conference*, 141–150
14. Islam M, Terao S and Murase K 2001, "Effect of Fitness for the Evolution of Autonomous Robots n an Open-Environment", *Proceedings of the 4th International Conference on Evolvable Systems*, Lecture Notes in Computer Science, Springer-Verlag, Berlin, 171–181
15. Islam M, Terao S and Murase K 2001, "Incremental Evolution of Autonomous Robots for a Complex Task", *Proceedings of the 4th International Conference on Evolvable Systems*, Lecture Notes in Computer Science, Springer-Verlag, Berlin, 181–191
16. Keymeulen D, Konaka K, and Iwata M 1997, "Robot Learning using Gate Level Evolvable Hardware", *Proceedings of the 6th European Workshop on Learning Robots*
17. Krohling R, Zhou Y and Tyrrell A M 2003, "Evolving FPGA-based robot controller using an evolutionary algorithm", *First International Conference on Artificial Immune Systems*, 41–46
18. Lala P K 1997 *Digital Circuit Testing and Testability*, Academic Press, San Diego
19. Lee P and Anderson T 1990, *Fault Tolerance Principles and Practice*, Springer-Verlag, 2nd Ed.
20. Levi D and Guccione S 1999, "Geneticfpga: Evolving stable circuits on mainstream FPGA devices" *Proceedings First NASA/DoD Workshop on Evolvable Hardware*
21. Levi D 2000, "HereBoy: A Fast Evolutionary Algorithm", *Proceedings 2nd NASA/DoD Evolvable Hardware Workshop*, 17–24
22. Liu H, Miller J and Tyrrell A 2005. "Intrinsic Evolvable Hardware Implementation of a Robust Biological Development Model for Digital Systems", *Proceedings 6th NASA/DoD Conference on Evolvable Hardware*, 87–92

23. Mange D, Sipper M, Stauffer A and Tempesti G 2000, "Towards Robust Integrated Circuits: The Embryonics Approach", *Proceedings of the IEEE* 88(4), 516–541
24. Metta G, Sandini G, Konczak J 1998, "A developmental approach to sensor-motor coordination in artificial systems", *IEEE International Conference on Systems, Man, and Cybernetics*, 3388–3393
25. Michel O 1999, *Khepera User Manual. K-TEAM*, Version 5.02
26. Miller J 2003, "Evolving developmental programs for adaptation, morphogenesis, and self-repair", *Seventh European Conference on Artificial Life*, Lecture Notes in Artificial Life, Vol. 2801, 256–265
27. Miller J and Thomson P 2000, "Cartesian genetic programming", *Lecture Notes in Computer Science*, Vol. 1802, 121–132
28. Miller J 2004, "Evolving a self-repairing, self-regulating, French flag organism", *Proceedings Genetic and Evolutionary Computation Conference*, 129–139
29. Nolfi S and Floreano D 2000, *Evolutionary Robotics: The Biology, Intelligence and Technology of Self-Organizing Machines* Cambridge, MA: MIT Press
30. Ortega C and Tyrrell A M 1998, "MUXTREE revisited: Embryonics as a Reconfiguration Strategy in Fault Tolerant Processor Arrays", *Lecture Notes in Computer Science*, Vol. 1478, Springer-Verlag, Berlin, 206–217
31. Sanchez E, Mange D, Sipper M, Tomassini M, Perez-Uribe A and Stauffer, A 1997, "Phylogeny, Ontogeny, and Epigenesis: Three Sources of Biological Inspiration for Softening Hardware", *Lecture Notes in Computer Science* Vol 1259, Springer-Verlag, Berlin, 35–54
32. Scholefield P 1960, "Shift Registers Generating Maximum-Length Sequences", *Electronic Technology* 10, 389–394
33. Schwefel H-P 1981, *Numerical Optimization of Computer Models*, Chichester: Wiley
34. Tan K C, Chew C M, Tan K K, Wang L F and Chen Y J 2002, "Autonomous Robot Navigation via Intrinsic Evolution" *Proceedings of the Congress on Evolutionary Computation 2002*, 1272–1277
35. Thompson A 1995, "Evolving Electronic Robot Controllers that Exploit Hardware Resources", *Proceedings of the 3rd European Conf. on Artificial Life*, Springer-Verlag, Berlin, 640–656
36. Thoma Y, Tempesti G, Sanchez E and Moreno J-M 2004, "POEtic: An Electronic Tissue for Bio-Inspired Cellular Applications", *BioSystems* 74(1–3), 191–200
37. Thoma Y, Sanchez E, Hetherington C, Roggen D, Moreno J-M 2004, "Prototyping with a bio-inspired reconfigurable chip", *Proceedings 15th International Workshop on Rapid System Prototyping*
38. Tyrrell A M, Hollingworth G S and Smith S L 2001, "Evolutionary Strategies and Intrinsic Fault Tolerance", *Proceedings of the 3rd NASA/DoD Workshop on Evolvable Hardware*, 98–106
39. Tyrrell A M, Krohling R A, and Zhou Y 2004 "A New Evolutionary Algorithm for the Promotion of Evolvable Hardware", *IEE Proceedings of Computers and Digital Techniques*, 151(4) 267–275
40. Tyrrell A M, Sanchez E, Floreano D, Tempesti G, Mange D, Moreno J M, Rosenberg J and Villa A E P 2003, "POEtic Tissue: An Integrated Architecture for Bio-Inspired Hardware", *Proceedings of 5th International Conference on Evolvable Systems*, Trondheim, 129–140

41. Wolpert L 2002, *Principles of Development* 2nd Ed., Oxford University Press
42. http://www.celoxica.com/
43. *Virtex Field Programmable Gate Arrays Data Book* Version 2.5, Xilinx Inc. (2001)
44. Xilinx 1999, *JBits* Version 2.0.1
45. Agapie A 2001, "Theoretical analysis of mutation-adaptive evolutionary algorithms", *Evolutionary Computation* 9(2), 127–146
46. Avizienis A 2004, "Towards systematic design of fault tolerant systems", *IEEE Computer* 30(4), 51–58
47. Belk C, Robinson J, Alexander M, Cooke W and Pavelitz S 1997, "Meteoroids and orbital debris: effects on spacecraft", *NASA Reference Bulletin* 1408
48. Burns A and Wellings A 2001, *Real-Time Systems and Programming Languages*, Addison-Wesley-Longmain, 3rd Edition
49. Dunn W 2002, *Practical Design of Safety-Critical Computer Systems*, Reliability Press
50. Greenwood G and Zhu Q 2001, "Convergence in evolutionary programs with self-adaptation", *Evolutionary Computation* 9(2), 147–158
51. Gallagher H and Vigraham S 2002, "A modified compact genetic algorithm for the intrinsic evolution of continuous time recurrent neural networks", in W. Langdon et al., (Eds.), *Proceedings GECCO 2002*, 163–170
52. Greenwood G, Ramsden E and Ahmed S 2003, "An empirical comparison of evolutionary algorithms for evolvable hardware with maximum time-to-reconfigure requirements", in Jason Lohn et al. (Eds.), *Proceedings 2003 NASA/DOD Conference On Evolvable Hardware*, 59–66
53. Greenwood G, Hunter D and Ramsden E 2004, "Fault recovery in linear systems via intrinsic evolution", *Proceedings 2004 NASA/DOD Conference On Evolvable Hardware*, 115–122
54. Greenwood G 2005, "On the practicality of using intrinsic reconfiguration for fault recovery", *IEEE Transactions on Evolutionary Computation* 9(4), 398–405
55. Hollingworth G, Smith S and Tyrrell A 2000, "The intrinsic evolution of Virtex devices through internet reconfigurable logic", *Proceedings of the 3rd International Conference on Evolvable Systems ICES 2000*, LNCS 1801 (Springer, Berlin), 72–79
56. Hughes H and Benedetto J 2003, "Radiation effects and hardening of MOS technology: devices and circuits", *IEEE Transactions on Nuclear Science* 50(3), 500–521
57. Keane M, Koza J and Streeter M 2002, "Automatic synthesis using genetic programming of an improved general-purpose controller for industrially representative plants", in A. Stoica (Ed.), *Proceedings of the 2002 NASA/DOD Conf. On Evolvable Hardware*, 113–122
58. Keymeulen D, Zebulum R, Jin Y and Stoica A 2000, "Fault tolerant evolvable hardware using field-programmable transistor arrays", *IEEE Transactions on Reliability* 49(3), 305–316
59. Linden D 2002, "Optimizing signal strength in-situ using an evolvable antenna system", *Proceedings of the 2002 NASA/DOD Conf. On Evolvable Hardware*, 147–151
60. Mange D, Sipper M, Stauffer A and Tempesti G 2000, "Embryonics: a new methodology for designing field programmable gate arrays with self-repair and self-replicating properties", *Proceedings of the IEEE* 88(4), 416–541

61. MIL-STD-882D, *Standard Practice for System Safety*, Department of Defense, 10 February 2000
62. NASA-STD-8719.7, Facility system safety guidebook, January 1998
63. NGST yardstick mission, NGST Monograph No. 1, Next Generation Space Telescope Project Study Office, Goddard Space Flight Center, 1999
64. Schnier T and Yao X 2003, "Using negative correlation to evolve fault tolerant circuits", *Proceedings 5th International Conference on Evolvable Systems (ICES 2003)*, 35–46
65. Sekanina L and Drabek V 2000, "Relation between fault tolerance and reconfiguration in cellular systems", *Proceedings of the 6th IEEE online testing workshop*, 25–30
66. Stoica A, Keymeulen D, Zebulum R, Thakoor A, Daud T, Klimeck G, Jin Y, Tawel R and Duong V 2000, "Evolution of analog circuits on field programmable transistor arrays", in Jason Lohn et al., (Eds.), *Proceedings of the Second NASA/DOD Workshop On Evolvable Hardware*, 99–108
67. Stoustrup J and Blondel V 2004, "fault tolerant control: a simultaneous stabilization result", *IEEE Transactions on Automatic Control* 49(2), 305–310
68. Streeter M, Keane M and Koza J 2002, "Routine duplication of post-2000 patented inventions by means of genetic programming", in J. Foster et al., (Eds.) *Genetic Programming: 5th European Conference, EuroGP 2002*, 26–36
69. Tai A, Alkalai L and Chau S 1999, "On-board preventive maintenance: a design-oriented analytic study for long-life applications", *Performance Evaluation* 35(3–4), 215–232
70. Wismer M 2003, "Steady-state operation of a high-voltage multiresonant converter in a high-temperature environment", *IEEE Transactions on Power Electronics* 18(3), 740–748
71. Zebulum R 2003, NASA JPL, private communication

CHAPTER 6

FUTURE WORK

Aims: *Chapter 1 identified two uses for EHW: circuit synthesis and circuit adaptation (with the latter oriented towards fault tolerant systems). Both areas are currently being rigorously investigated by the EHW community. But does that mean both areas should be investigated? And if so, in what specific areas? In this chapter some answers are provided.*

6.1 CIRCUIT SYNTHESIS TOPICS

Circuit synthesis starts with a design specification with explicitly stated performance requirements. The EHW objective is to evolve a circuit that meets all of those specifications. In what follows is a description of the current state of the art in digital and analog circuit design and how EHW techniques compare. This comparison will show two things: (1) some research areas currently being investigated by the EHW community should be dropped, and (2) some research areas currently being ignored by the EHW community should be investigated.

Before starting, however, it is important to understand the purpose of EHW is to evolve circuit *behaviors* and not circuit *structures*. An essential ingredient of any EHW technique is an evolutionary algorithm that relies on a fitness function to evaluate a circuit. But what is evaluated is whether or not the circuit has the right frequency response or the right timing. These are behavioral properties. Circuit design always begins with a specification. Remember that the purpose of a design specification is to state *what* needs to be designed and not *how* to do

---

Introduction to Evolvable Hardware: A Practical Guide for Designing Self-Adaptive Systems,
by Garrison W. Greenwood and Andrew M. Tyrrell
Copyright © 2007 Institute of Electrical and Electronics Engineers

that design. In other words, the specification tells the designer what performance is required—it is up to the creativity of the designer to come up with a circuit that meets those requirements. Hence, EHW is a design method more in line with the way designers actual do their job.

6.1.1 Digital Design

Despite the enormous amount of digital design work done by industry every day, this is one area where the EHW community has had limited success breaking into. Much of the digital design work is automated, and yet no commercial digital design product boasts that it uses an EHW-based design method. To fully appreciate why this is so, it is important to understand how industry does digital design.

The overwhelming majority of digital design work done today uses a variety of electronic design automation (EDA) tools. Complicated designs are usually implemented in an FPGA or in an ASIC where the device density permits high-speeds in small packages. The complexity of these designs makes hand design methods impossible. EDA tools automate the design process thereby reducing design time, improving yields, and reducing non-recurring costs. Figure 6.1 shows the design process steps.

The process begins by describing the digital circuit in a computer program written in a *hardware description language* (HDL). The two most ubiquitous HDLs are Verilog and VHDL, both of which are specified by IEEE standards[1]. This design can be expressed in a mixture of different levels of abstraction. The compiled source code is input to a synthesizer along with a component library and any design constraints on timing or power consumption. The synthesizer takes the design described by the HDL and, using devices from the component library, creates a circuit that satisfies any design constraints. The synthesizer output feeds another EDA tool that assigns logic functions to configurable logic blocks inside an FPGA and determines the routing between blocks. A bitstream produced by the design implementation tool is used to physically program the FPGA.

Design verification occurs at various places in this design process. HDL descriptions are simulated to verify the design meets all functional specifications. A second simulation is performed after synthesis to ensure the synthesized circuit still functions properly. Once the synthesized circuit design is placed and routed a thorough timing analysis is conducted. Finally, the programmed FPGA is placed into its target system for a full functional and timing check. If the verification fails at any stage, the original HDL description must be modified and the synthesis process repeated. For example, if the timing analysis identifies a flaw, the designer could describe the design at a lower level of abstraction, which allows for a tighter control over what gets synthesized.

EHW practitioners must understand the FPGA design flow shown in Figure 6.1 is in place and widely used throughout the integrated circuit industry today. In

---
[1] IEEE Standard 1364-2001 covers Verilog; IEEE Standard 1076.6-2004 covers VHDL.

CIRCUIT SYNTHESIS TOPICS 179

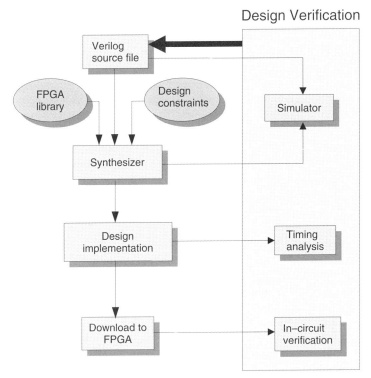

Figure 6.1. The FPGA design flow. Verilog is shown as the HDL, but VHDL could also be used. All tasks in the shading box perform design verification. The red arrow near the top means any failure during the verification must be handled by modifying the Verilog source code and then resynthesizing the design. Although FPGA synthesis is shown, the procedure is identical for ASIC design if the FPGA component library is replaced with a standard cell library.

essence, this means the EHW community has to show substantial, significant advantages over an established method before making any real inroads—and that presents several challenges to any EHW method trying to perform circuit synthesis. Indeed, an EHW method has to contend with the following:

- **Compatibility with EDA Tools**

    Any EHW-based method will have to work in the existing FPGA design environment. EDA tools used in this environment are available from a variety of different vendors. What makes this all work is standard file formats are used to ensure compatibility. For example, most synthesizers expect design descriptions inputs to be in the Electronic Design Interchange Format (EDIF), which is owned by the Electronic Industries Alliance (EIA)[2].

[2]EDIF Version 4 0 0 is now ANSI/EIA Standard 682-1996.

Many of the powerful analysis and verification techniques so critical to producing a successful design can't be used to evaluate evolved designs until the synthesis is completed. This means the EHW-based method must provide configurations in appropriate file formats such as EDIF. At present this isn't being done.

- **Handling Design Constraints**

  Virtually all FPGA designs must accommodate constraints on timing, power consumption and area. Consequently, any EHW-based method, at a minimum, must solve a constrained optimization problem, while in the worst case it must solve a multiobjective optimization problem. Unfortunately, Pareto searches aren't practical because they won't present a single solution to the synthesizer. Moreover, this puts the designer in the middle of the FPGA design process. (The process shown in Figure 6.1 is completely automated once the HDL program is written.) To make matters even worse, verifying if a design constraint is met requires using commercial EDA analysis tools, which is another reason why file formats are important.

- **Coping with Verification Problems**

  Designs implemented in FPGAs are naturally complex. Verification is absolutely essential at various stages in the design flow process to ensure specifications are met. The only thing a designer can do when a verification test fails is to modify the HDL source code. As mentioned above, going to a lower level of abstraction in the HDL program can help meet a timing constraint. With EHW techniques the only recourse is to run the EA longer.

- **Scalability**

  Scalability has been a real issue for the EHW hardware community for some time (e.g., see [10, 5, 9]) and the situation has not improved a whole lot. The problem is simple: as the design complexity grows the genotype size lengthens. Any genotype increase causes an exponential growth in the search space, thereby dramatically increasing the search time. No good solution to this problem exists, although some recent approaches such as *morphogenesis* show some promise. The main idea behind morphogenesis is to have the genotype encode a growth process rather than an explicit circuit configuration [6]. That way the complexity is reflected in the genotype-phenotype mapping rather than in the genotype itself. But, the circuitry in which morphogenesis has been tried so far is relatively simple.

  To be perfectly blunt, we must paint a very pessimistic future for EHW-based digital circuit synthesis efforts. In reality the EHW-based approaches simply cannot compete—and probably never will be able to compete—with existing FPGA design methods. Take another look at Figure 6.1 and consider that FPGA library block. All FPGA vendors furnish pre-defined modules, of amazing complexity, for that FPGA library. Synopsys, for example, sells a library of synthesizable Verilog and VHDL modules in its DesignWare® Intellectual Property product line. These modules range from simple multiplexers to flash memory controllers to complete USB 2.0 Host Controllers [4]. Nothing currently published in the EHW literature can even approach this level of complexity.

But complex functions are not the only advantage these pre-defined modules provide. Many modules are specifically designed to exploit the internal architecture of the FPGA. For instance, the Xilinx SRL16 library module configures internal look-up table memory in its FPGAs to act like a 16-bit shift register without using the flip-flops available in each slice [8]. Any reduction in needed resources helps to meet area constraints. The problem for the EHW approach is how to tradeoff exploration, which searches for a configuration to implement a function, with exploitation, which searches for efficient resource usage. The pre-defined library modules have already done this tradeoff. This tradeoff still plagues EAs [1].

Arguably the scalability issue is the most difficult obstacle the EHW community has to overcome. Scalability problems have really hampered efforts at evolving more complex designs. No real work on compatibility with EDA tools or competition with commercial Intellectual Property product lines can really begin until scalability has been resolved.

Design-for-test (DFT) is perhaps the one area in digital circuit synthesis field where EHW can make a positive contribution. For some unknown reason this area has been completely ignored by the EHW community. The purpose of DFT is to put "hardware hooks" into a circuit so that quality and reliability measurements may be taken [2]. In other words, DFT involves modifying an existing design solely to assist in fault detection and isolation. This additional circuitry is only used for testing and should have no effect on the normal circuit operation. Indeed, the circuit must be put into a "test mode" to activate this testing circuitry. An example will illustrate the main idea.

Figure 6.2 shows a combinational logic circuit that implements $Y = \overline{A}B + AC$. Table 6.1 shows the truth table of this circuit. It also shows how the circuit should respond during an exhaustive test[3]. Suppose one of the gates has its output accidently connected to logic ground, which is called a *stuck-at-0 fault*. An exhaustive test would identify this fault. For instance, if the U2B gate output has a stuck-at-0 fault, then the inputs $ABC = 101$ and $ABC = 111$ would make Y logic 0 instead of logic 1.

The original circuit, however, does have a design flaw. An input transition $111 \rightarrow 011$ should not cause Y to change, but in fact there is a glitch, called a hazard, where Y briefly goes to logic 0. These hazards arise whenever there are unequal propagation delay paths in the circuit. The solution is to add a hazard cover. This cover is an added gate, which also adds another minterm (BC) to the

---

[3]An exhaustive test evaluates the circuit performance using every possible input.

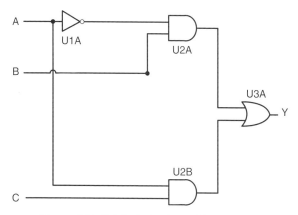

Figure 6.2. Original circuit used in example.

TABLE 6.1. Truth Table for the Example Problem

A	B	C	Y
0	0	0	0
0	0	1	0
0	1	0	1
0	1	1	1
1	0	0	0
1	0	1	1
1	1	0	0
1	1	1	1

boolean expression, to compensate for the unequal delay paths. Figure 6.3 shows the modified circuit. It is easy to verify the truth table still holds.

A stuck-at-one fault on the output of any gate would show up during an exhaustive test. Most stuck-at-0 faults would as well—with one important exception. Unfortunately, a stuck-at-0 fault on the output of U2C cannot be detected. (The reader should verify this.) However, by adding some additional circuitry it will be possible to detect such a fault. Figure 6.4 shows the modification. Notice there is now an additional gate (U2D) and another input *TEST*. When $TEST = 1$, the circuit operates as normal. However, $TEST = 0$ effectively removes the $\overline{A}B$ minterm from the boolean expression. Of course, removing a minterm changes the results of an exhaustive test (see Table 6.2). Even while in test mode the modified circuit should have $Y = 1$ for a 011 or a 111 input. But now a stuck-at-0 fault on the U2C output is detectable because that fault makes $Y = 0$ in the modified circuit. (It would still be logic 1 when not in test mode.)

The above example, although simple, does illustrate why DFT is so important: exhaustive testing alone is not guaranteed to find all possible faults. However, when exhaustive testing is used in conjunction with a small amount of additional

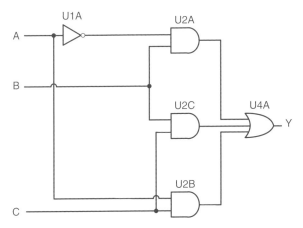

Figure 6.3. Modified circuit with the hazard cover U2C.

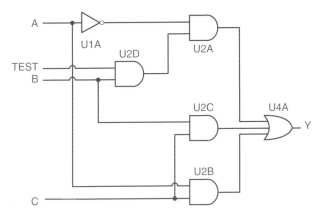

Figure 6.4. Final circuit modified with a test gate U2D. The circuit is in test mode when TEST = 0.

circuitry, the likelihood of detecting a fault improves dramatically. Unfortunately incorporating DFT is not always as easy as the simple example above might suggest. Indeed, in most cases the exact circuitry to add, and where it should be placed, will not be obvious. This is where EHW can play a role by evolving the test circuitry.

It is important to emphasize evolving test circuitry to insert into an existing circuit will not be easy. First of all the additional circuitry cannot alter the functionality of the original circuit when that circuit is in normal operational mode. (In test mode the original circuit functions may not work, but that is okay because the purpose of test mode is to identify and isolate faults.) An unavoidable consequence of inserting test circuitry is it will add delays. Hence, each evolved test circuit

TABLE 6.2. Truth Table for the Circuit with DFT Modifications

TEST	A	B	C	Y
0	0	0	0	0
0	0	0	1	0
0	0	1	0	0
0	0	1	1	1 ⇐
0	1	0	0	0
0	1	0	1	1
0	1	1	0	0
0	1	1	1	1
1	0	0	0	0
1	0	0	1	0
1	0	1	0	1
1	0	1	1	1
1	1	0	0	0
1	1	0	1	1
1	1	1	0	0
1	1	1	1	1

Notice the output is identical to Table 6.1 when $TEST = 1$. The highlighted line has $Y = 0$ if the hazard cover gate has a stuck-at-0 fault.

must be checked for feasibility—that is, it doesn't violate any functional or timing specification in the original circuit. This will prove to be a very difficult constrained optimization problem. Nevertheless, EHW could be of enormous assistance to the designer because currently this process is done by hand.

6.1.2 Analog Design

In contrast to the digital design field, analog circuit design is a wide-open field where EHW-based methods can make significant contributions. Indeed, the EDA environment for analog design is nowhere near as mature as it is for digital design. HDLs for analog design do exist; Verilog-A and Verilog-AMS are the most prominent. However, these HDLs are currently used only for simulation.

It is instructive to see how FPAA vendors approach analog design. At present designers use graphical interfaces to communicate with vendor-supplied software, but that software doesn't synthesize anything in the conventional sense. Lattice Semiconductor, for example, provides PAC-Designer™ software for its ispPAC devices. This software provides pull-down menus for selecting clock signaling specifications, and "point-and-click" schematic entry for setting internal connections and choosing parameter values. For some devices Lattice provides a database of pre-designed blocks. SPICE model export is provided to support simulation [7].

Most practitioners fix the circuit topology and only concentrate on choosing component values. This approach reduces the search space, but at the same time this approach limits the ability to explore alternative circuit configurations that might produce more efficient designs or novel performance. A thorough analysis of the difficulties surrounding evolving analog circuit topologies would help develop efficient design methods. This is an open area for investigation.

6.2 CIRCUIT ADAPTION TOPICS

This is the area where EHW methods have the most interest and the greatest potential for both digital, analog and mixed signal systems. Circuitry can be adapted—that is, reconfigured—to take on new roles or for fault recovery. New roles are relatively easy to accommodate if those roles are known beforehand because new reconfigurations can always be predefined and then stored until needed. Consequently, the focus in this section is on adaptation for fault recovery operations.

Circuits are adapted *in-situ*. What makes such an environment particularly difficult to work in is the user almost never has complete knowledge about why the original circuitry failed. Obviously faults can degrade a circuit's performance, but any change in the operational environment can do this as well. Regardless of the cause of a fault, reconfiguration done *in-situ* is especially challenging for two reasons: (1) faults can be hard to detect and isolate, and (2) the reconfiguration function itself may be compromised by the fault. Poor fault isolation means the user really doesn't know what circuitry still remains operational. This makes it impossible to know how much (if any) functionality can be recovered or how long recovery operations can take. The second reason is even worse because, without a fully functional reconfiguration capability, it may not be possible to recover from the fault—even if the required resources are available and fully operational. With this in mind, the following areas are open to investigation:

- **Intrinsic vs. Extrinsic Evolution**

    Extrinsic evolution is impractical for *in-situ* reconfiguration because software models can't be depended upon to accurately capture a faulty system's dynamics or a changed operational environment. Evolution under these circumstances is futile.

    At the present time a significant portion of EHW-based fault recovery investigations rely on simulations. While this approach may be suitable for developing and optimizing an EA, the actual tests to measure EA efficacy have to be done intrinsically. Most intrinsic evolutions are currently done in laboratory environments where spectrum analyzers, digital storage oscilloscopes and other instruments are available.

    Investigations are needed to develop intrinsic recovery methods for autonomous systems with limited measurement resources.

- **Restricted Functionality**

  EHW-based fault recovery investigations usually assume the reconfigurable circuitry can, given an effective EA and a sufficient running time, completely restore any lost functionality. However, serious faults or extreme environmental changes may degrade a significant portion of the circuit to a point where complete restoration of lost functions isn't possible. This degradation may prevent some circuit configurations from even being constructed. In effect, entire regions of the fitness landscape are now no longer reachable. Extrinsic evolution is impractical without precise knowledge of what can and what cannot be configured. Hence, only intrinsic evolution is useful here.

  More work needs to be done on developing EAs that can intrinsically evolve circuit configurations with imprecisely defined performance objectives. (Imprecise in the sense one does not know what level of performance can be achieved.) One possible approach is to explore intrinsic evolution in stages where each stage has a different performance objective. Evolving benign configurations where further damage is contained and controlled is also of interest.

- **Degraded Reconfiguration Resources**

  Most EHW-based recovery methods work with fully functional reconfigurable resources—a presumption that may not be justified. For example, it is known that high radiation environments cause device characteristics to change [3]. Faults induced by high radiation are just as likely to degrade FPAAs, FPTAs, and FPGAs as any other circuitry. Moreover, the *in-situ* computing resources that run the EA are also subjected to the same operational environment, which makes them just as susceptible to high radiation effects.

  Studies are needed to determine how effective EHW-based recovery methods are when the computing resources they run on are degraded by environmental conditions.

- **Reconfiguration Platforms**

  Most EHW investigations are conducted in laboratory environments where there are no restrictions on the computing resources needed to run the EA. Unfortunately, *in-situ* environments never have this luxury.

  Investigations are needed to develop EAs that run on small embedded systems where computing speeds and memory availability is more in line with what is available on deployed, autonomous systems. These conditions will permit more realistic assessments of EHW recovery methods particularly under tight recovery deadlines.

- **Testing Scenarios**

  Many fault recovery scenarios involve injecting arbitrary faults into an operating circuit—for instance, a randomly chosen switch in an FPTA is forced open. It is not clear if such a fault is likely to occur in isolation or whether it results from some other fault. This ambiguity leads to the development of recovery methods that may have limited usefulness.

Investigations are needed where an FMEA identifies the faults and then an FMET evaluates the EHW-based recovery method. Any faults or sequence of faults used during the FMET are restricted to those identified in the FMEA.

- **System Status During Reconfiguration**

  Unfortunately, circuits do not operate in isolation. Any circuit undergoing recovery operations—whether via EHW-based methods or not—is going to have some effect on any circuitry it interfaces to. Most EHW fault recovery work is conducted on independent circuits where any interfaced circuitry can be ignored. This isn't realistic.

  EAs must be expected to perform reconfiguration without causing additional faults to be injected into non-faulty interfaced circuits. Studies are needed to develop ways to evolve circuitry while keeping all other system interfaces intact.

EHW techniques can dramatically improve the reliability and availability of deployed hardware, with minimal interruptions to their operations. Its now time to start applying these techniques to real-world problems. As stated earlier in this chapter, digital design is never going to generate a whole lot of interest outside the EHW community, but adaptive systems has an almost unlimited potential. Autonomous systems such as deep space probes are likely applications, which explains why NASA is so interested in EHW. There are even military applications for EHW. For instance, underwater autonomous vehicles for the U.S. Navy is an exciting new application area. These vehicles operate for extended periods of time in harsh, undersea environments with little or no human intervention. Fault tolerance and adaptability are key to their survival.

The promises and challenges of EHW are known. Hopefully this book will inspire the reader to learn more about this exciting field.

REFERENCES

1. Eiben A and Schippers C 1998, "On evolutionary exploration and exploitation", *Fundamenta Informaticae* 35, 35–50
2. Crouch A 1999, *Design for Test for Digital IC's and Embedded Core Systems*, Prentice-Hall, Upper Saddle River, NJ
3. Hughes H and Benedetto J 2003, "Radiation effects and hardening of MOS technology: devices and circuits", *IEEE Transactions on Nuclear Science* 50(3), 500–521
4. See http://www.synopsys.com/designware/ for details
5. Haddow P, Tufte G and Remortel P 2001, "Shrinking the genotype: L-systems for EHW?", *Proceedings of the International Conference on Evolvable Systems*, 128–139

6. Lee J and Sitte J 2005, "Issues in the scalability of gate-level morphogenetic evolvable hardware", in Abbass, Bossamaier and Wiles (Eds.), *Recent advances in artificial life*, 145–158
7. Lattice Semiconductor Appl. Note AN6021 2000, *PSpice Simulation Using ispPAC SPICE Models and PAC-Designer*
8. Xilinx Appl. Note XAPP465 (v1.0) 2003, *Using Look-UP Tables as Shift Registers (SRL16) in Spartan-3 Devices*
9. Vassilev V and Miller J 2000, "Scalability problems of digital circuit evolution: evolvability and efficient designs", *Proceedings 2nd NASA/DOD Workshop on Evolvable Hardware*, 55–65
10. Yao X and Higuchi T 1999, "Promises and challenges of evolvable hardware", *IEEE Transactions on Systems, Man & Cybernetics—Part C* 29(1), 87–97

INDEX

adaptive hardware, 15, 101
adaptive systems, 102, 189
aging effects, 16, 112, 121, 123, 126–127, 144
apoptosis, 153
artificial immune system, 143
ASIC, 35, 54, 156, 182

cartesian genetic programming, 49, 53, 157
circuit synthesis, 15, 181
   analog, 188
   digital, 182
correlation, 115
COTS, 57, 59, 68, 96, 98, 110, 123, 128–129, 176
CPLD, 38

design
   analog, 188
   digital, 182
design complexity, 184
design-for-test, 185
directed graph (digraph), 22, 49

electronic design automation, 182, 184, 185
embryonic, 17, 143
   array, 147
   cell, 151, 155

embryonic array, 149
event
   anticipated, 116
   unanticipated, 116
evolutionary algorithm
   evaluation, 25
   fitness, 19
   fitness landscape, 26–27, 173
   getting it to work, 31
   multiobjective, 29
   representation, 20
   selection, 28
   steady-state, 30, 52
   termination, 30
   variation, 22
evolvable hardware, 11
evolved hardware, 11
exogenous events, 116
exploitation, 32, 185
exploration, 32, 185
extrinsic evolution, 13, 15, 75, 102, 189

failure, 103
failure mode, 105
fault, 103
   anticipated, 106, 114, 117
   detection, 103, 147, 150

fault (*continued*)
    recovery, 103, 105, 189
    unanticipated, 106, 114, 116
fault tolerant system, 102
    fitness-based design, 118
    population-based design, 118
first principle of fault recovery, 112, 117–118
FMEA, 105, 112, 114
FMET, 105
FPAA, 74, 188
FPGA, 36, 38, 184
FPTA, 74, 109, 117–118
    Heidelberg device, 87
    NASA device, 74
FTA, 105, 112, 114

graph, 22

hazard, 103

in-situ, 15, 109, 114, 117, 189
intrinsic evolution, 13, 15, 49, 102, 189

Khepera robot, 11, 129, 132, 140

logically correct, 106, 112–115

mishap, 103
mixtrinsic evolution, 85

morphogenesis, 184

offline evolution, 15
online evolution, 15

Pareto optimal, 29, 184
PLD, 36, 37
POEtic, 59, 143–147, 155
POEticmol, 155
preventive maintenance, 115

real-time system, 106, 112, 134
reconfigurable hardware, 101
reconfiguration, 16, 39, 189–191
    extrinsic, 108, 190
    intrinsic, 108
redundancy, 105, 115, 147, 149, 151
risk, 16
    minimizing, 114, 116
    mishap, 103

SABLES, 80, 82, 85
scalability, 184–185
synthesis, 101

temporally correct, 106–107, 112–113
tree, 22, 49

undirected graph, 22

ABOUT THE AUTHORS

Andy Tyrrell received a first class honors degree in 1982 and a PhD in 1985, both in Electrical and Electronic Engineering. He joined the Electronics Department at York University in Heslington, York, United Kingdom in April 1990 and was promoted to Chair in Digital Electronics in 1998. Previous to that he was a Senior Lecturer at Coventry Polytechnic. Between August 1987 and August 1988 he was visiting research Fellow at Ecole Polytechnic in Lausanne, Switzerland where he researched into the evaluation and performance of multiprocessor systems. From September 1973 to September 1979 he worked for STC at Paignton Devon on the design and development of high frequency devices.

His main research interests are in the design of biologically-inspired architectures, artificial immune systems, evolvable hardware, FPGA system design, parallel systems, fault tolerant design, and real-time systems. In particular, over the last six years his research group at York has concentrated on bio-inspired systems. This work has included the creation of embryonic processing array, intrinsic evolvable hardware systems, and the immunotronics hardware architecture. He is Head of the Intelligent Systems research group at York, as well as Head of Department. He was General Program Chair for ICES 2003 and Program Chair for IPCAT 2005. He is a both a Senior member and a Fellow of the IEEE.

Garrison Greenwood received a PhD in Electrical Engineering from the University of Washington, Seattle. After spending more than a decade in the industry designing multiprocessor embedded system hardware, he entered

---

Introduction to Evolvable Hardware: A Practical Guide for Designing Self-Adaptive Systems, by Garrison W. Greenwood and Andrew M. Tyrrell
Copyright © 2007 Institute of Electrical and Electronics Engineers

academia where he is now an Associate Professor in the Department of Electrical and Computer Engineering at Portland State University, Portland, OR. From 1999 to 2000 he was a National Science Foundation Scholar-in-Residence at the National Institutes of Health.

His research interests are evolvable hardware and adaptive systems particularly when used in fault tolerant applications. Dr. Greenwood is a member of Tau Beta Pi, Eta Kappa Nu, a Senior member of the IEEE, and is a registered Professional Engineer. Since 2000 he has been an Associate Editor of the IEEE TRANSACTIONS ON EVOLUTIONARY COMPUTATION and currently serves as Vice President (Conferences) for the IEEE Computational Intelligence Society.